能量与热力学建筑书系
Energy & Thermodynamic Architecture

热力学建筑视野下的空气提案
设计应对雾霾

同济大学人居环境生态与节能联合研究中心资助（项目名称：能量形式化与热力学建筑前沿理论建构）
国家自然科学基金资助（项目批准号 51278340）

Air through the Lens of Thermodynamic Architecture
DESIGN AGAINST SMOG

李麟学　周渐佳　谭峥　著
Li Linxue　Zhou Jianjia　Tan zheng

同济大学出版社
TONGJI UNIVERSITY PRESS

图书在版编目（CIP）数据

热力学建筑视野下的空气提案：设计应对雾霾／李麟学，周渐佳，谭峥著 . -- 上海：同济大学出版社，2015.12
（能量与热力学建筑书系／李麟学主编）
ISBN 978-7-5608-6150-0

Ⅰ.①热... Ⅱ.①李... ②周... ③谭... Ⅲ.①热力学
－应用－建筑设计 Ⅳ.① TU2

中国版本图书馆 CIP 数据核字（2015）第 320144 号

同济大学人居环境生态与节能联合研究中心资助（项目名称：能量形式化与热力学建筑前沿理论建构）
国家自然科学基金资助（项目批准号 51278340）

热力学建筑视野下的空气提案
设计应对雾霾

李麟学 周渐佳 谭 峥 著

出 品 人：	华春荣
策 划：	秦蕾／群岛工作室
责任编辑：	晁 艳
特约编辑：	王轶群
责任校对：	徐逢乔
平面设计：	张 微
	设计绵延工作室
版 次：	2015 年 12 月第 1 版
印 次：	2015 年 12 月第 1 次印刷
印 刷：	上海安兴汇东纸业有限公司
开 本：	889mm×1194mm 1/16
印 张：	14.5
字 数：	290 000
书 号：	ISBN 978-7-5608-6150-0
定 价：	89.00 元
出版发行：	同济大学出版社
地 址：	上海市四平路 1239 号
邮政编码：	200092
网 址：	http://www.tongjipress.com.cn
经 销：	全国各地新华书店

本书若有印装问题，请向本社发行部调换
版权所有 侵权必究

Air through the Lens of Thermodynamic Architecture:
DESIGN AGAINST SMOG

by: LI Linxue, ZHOU Jianjia, TAN Zhen

ISBN 978-7-5608-6150-0

Initiated by: QIN Lei / Studio Archipelago

Produced by: HUA Chunrong (publisher),
CHAO Yan , WANG Yiqun(editing), XU Fengqiao
(proofreading), ZHANG Wei (graphic design)

Published in December 2015, by Tongji University Press,
1239, Siping Road, Shanghai, China, 200092.
www.tongjipress.com.cn

All rights reserved
No part of this book may be reproduced in any
manner whatsoever without written permission from
the publisher, except in the context of reviews.

序言

伊纳吉·阿巴罗斯

"设计应对雾霾"是个令人兴奋的标题，它使我们即刻开始重新思考设计在当代城市中扮演的角色。"热力学建筑视野下的空气提案"更是恰到好处地打消了人们最后的疑虑：这是在热力学原理基础之上设计空气，或是用空气进行设计。

李麟学教授为同济大学国际暑期课程提出的方案直截了当，激动人心。一些来自世界各地最权威建筑学校的优秀学生受邀，重新思考那个与时间密切相关，被现代人粗略定义为"空间"的概念，正在向物质性方向转变：空气是挥发性材料，而这种不稳定的物质，不过是由微小的颗粒构成。因此对它们的处理，可以通过物理和化学原理得以完成，也就是当代科学所描述的，能够最终决定生死和时间的热力学原理。

因此，"设计应对雾霾"可以说是一次从非材料化的情形出发，与生活和财富息息相关的设计邀请。这一方法对于上海来说尤为重要；在中国，热力学理论的应用已经不再是假设性的建议，而是与社会政治生活紧密关联的事件。如今，形态恢宏，装饰独特的花纹或肌理不再意味着建筑的品质。恰恰相反，品质存在于一种强烈、愉悦而多样的生活体验，以及一座将人们日常生活变得坦诚而亲切的城市中。

"中国城市向新常态回归"口号的提出，向保守主义和怀旧之情打开了大门。在对"旧日美好时光"的渴望和呼唤中，许多人从中受益。常态——假如如今还存在所谓常态的话，只能通过现实生活质量进行表达，并在物质文化中得以呈现。我们周围的物质，支持着我们的日常活动，并使它们成为一条引领人类进步的康庄大道。

对李麟学教授来说，热力学意味着更好的生活。既是一种建筑特质的别样呈现方式，又是一条建筑教学的独到路径，更是一种注视学科过去并对其现代性进行评价的方法。同时也是面对未来，寻找我们的科学知识何去何从的另一种方式，当我们在一个日新月异的城市发展环境中进行设计，城市发展速度和尺度已经将一切回头路和怀旧可能性断绝，既非过去，也非现代。我们就像一位孤儿失去了双亲，孑然而立，面临着没有传统也没有现代的矛盾二元对立状态。但是现在我们开始知道如何处理这种建筑材料，如何用空气建造我们的建筑。这种物质包含了一切潜力和维度，既是答案又是问题。

诚如所见，工作坊主题的提出及参与者的遴选非常高瞻远瞩。但是，用这种快速的学术模块化的暑期工作坊回应如此艰深的跨学科问题是否可能？手捧本书的读者也有权利质疑。作为一个聆听并广泛讨论过各个展出团队作品的评委会成员，我的观点是完全肯定的。学生们的热情、高水准和对知识的渴望，加上李麟学教授炉火纯青的团队管理技巧，催生了工作坊和著作的雏形。读者们也可以提前体会到未来时代的真实示例。在我看来的确如此，极富竞争力的一代人构想出可以象征室内和室外空间特质的设计形态，并且重新思考在新的公共边界中如何构建建筑类型和街道生活。这一切创意的集合表明，将热力学应用于设计是一种强大的工具，可以用其重新定义城市规划和设计中的失败过程，以及仍被人们广泛采用的技术。

Preface

Iñaki Abalos

DESIGN AGAINST SMOG is a provocative headline. It makes us immediately rethink the role of Design in Contemporary City. AIR THROUGH THE LENDS OF THERMODYNAMIC ARCHITECTURE, dispels any last doubt: It is about designing air or with air on a Thermodynamic Principles basis. No more, no less.

Professor Li Linxue's proposal for Tongji University International Summer School can't be any more thrusting or pertinent. What is he suggesting? A number of chosen students from some of the most prestigious Architecture Schools are invited to rethink that bastardly defined thing moderns called Space, subtly related to Time, now turned into a materially qualified something—Air, that volatile material, constituted however of particles and corpuscles, whose handling may attend to physic and chemical principles, described by contemporary sciences as the thermodynamic principles governing life and death, time in the end.

DESIGN AGAINST SMOG is therefore an invitation to broach Design from its most immaterial condition, those directly linked to Life and Wealth if I dare say. The approach is specially relevant to be developed in Shanghai and China, where the application of scientific thermodynamic criteria is no longer considered an abstract intellectual hypothesis, but a basic social and political matter. This days spectacular shape, ornamental pattern or textural frenzy, does not mean quality anymore. Very otherwise, quality relies in the pure experience of a differential way of life, intensified and pleasant, in a not so merciless city asked to become conspiratorial with people daily life.

"The return to a new normality in Chinese cities" has been spoken, and many have taken profit of this open door to conservatism and nostalgia, in a longing cry for "better past times". Normality, if something like that still exists nowadays, could only be expressed in real-life quality terms, and throughout a material culture. Materials surround us, support our quotidian activities, allow them to be a dignified real path to progress.

To Professor Li Linxue, Thermodynamics means better life. An other architectonic quality, other way of teaching architecture, of gazing the past and judging Modernity. Other way to face the Future and find the place our scientific knowledge drives to, while designing in a context of ever faster urban growing, already reached speeds and scales that make unfeasible any way back, any nostalgic reaction, nor this of the old neither that of the moderns. Orphans, parentless children. Left alone, with no tradition or modernity, we face a paradoxical duality. But now we start to know how to handle this building material, how to make our buildings out of Air. Both the solution and the problem, the material contents all the potentialities and dimensions.

As we already said, the posed subject of the workshop and the participants selection is as ambitious as it can be. But, is it possible to respond to all this deep cross-disciplinary questions through this fast-kind academic modules summer workshops are? The reader that holds this book in his hands has the right to answer himself back. My opinion as a Jury Member, who listened and largely discussed the exhibited team's work, cannot be other than positive. Passion, quality, and thirst for knowledge, lent on Professor Li Linxue's team managing adroitness, gave birth to embryonic workshop and book edition. The reader can taste what in my opinion is a unique true example of a future time to come given in advance. An extremely competent generation figures out the designed shapes that may embody interior and exterior spatial quality, and rethinks building typologies and street life within new public boundaries. This whole ensemble of ideas demonstrate that thermodynamics applied to design is a powerful tool to redefine failed processes and techniques some keep on working within urban planning and design.

能量与热力学建筑书系

《设计应对雾霾——热力学建筑视野下的空气提案》

《形式追随能量——热力学建筑设计前沿》

《热力学建筑原型》

《热力学建筑唯物论》

《建筑自然系统论》

《自然系统建构——麟和建筑设计实践》

目录

005　**序言** ○ 伊纳吉·阿巴罗斯

主题文章

014　**为何研究空气？** ○ 李麟学
033　**捕捉空气：空气与建筑的几种可能** ○ 周渐佳
041　**《北京蓝》的影像叙事** ○ 王子耕
047　**应对慢性灾害：烟尘、废气和雾霾** ○ 艾奥里奥·索托·汉德拉纳塔
063　**新邻里模型：新城市主义的热力学** ○ 谭峥
077　**建筑废料纪念碑** ○ 刘宇扬
087　**步骤：对社区可持续性问题的系统性思考** ○ 苏运升，乔治·吉奥吉夫
093　**城市地区空气污染现状和研究方法** ○ 李卓，陶文铨

设计成果

106　非"异形"热力学
122　不可见的基础设施
136　净化：生态介入
150　核：空气基础设施
164　热力学呼吸系统
178　藤
194　步骤：退出雾霾
208　互惠的城市主义

附录

Energy & Thermodynamic
Architecture Book Series

Air through the Lens of Thermodynamic
Architecture: DESIGN AGAINST SMOG

Form Follows Energy: Frontier for Thermodynamic
Architectural Design

Thermodynamic Architectural Prototypes

Thermodynamic Architectural Materialism

Theory on Natural System in Architecture

Construction of Natural System: Architectural
Practice of Atelier L+

Contents

005 Preface/Iñaki Abalos

ESSAYS

014 **Why Air?** /Li Linxue
033 **Catching Air: The Possibilities of Air and Architecture** /Zhou Jianjia
041 **Visual Narration of *Beijing Blue*** /Wang Zigeng
047 **Engaging with Slow Catastrophe: Smoke, Gas, and Smog** / Erioseto Hendranata
063 **The New Neighborhood Model: The Thermodynamics of New Urbanism**/Tan Zheng
077 **Memorial of Building Remains** / Liu Yuyang
087 **STEP: A Systematic Consideration of Settlement Sustainability Issues**/ Su Yunsheng, Georgi Georgiev
093 **Air Pollution Status in Urban Areas and Research Methods** /Li Zhuo, Tao Wenquan

WORKS

106 DeMONSTERative Thermodynamics
122 Invisible Infrasturcture
136 Purify: Ecological Intervention
150 Core: Air Infrastructure
164 A Thermodynamically Driven Respiratory System
178 The Vine
194 STEP: Esc Smog
206 RECY-Procal Urbanisms

APPENDIX

主题
文章

ESSAYS

设计应对雾霾 DESIGN AGAINST SMOG

图 1. 雾霾下的北京与上海 CBD 区域

为何研究
空气？

李麟学

什么是空气？一个巨大的虚空，无物。

世界上所有不可见层的透明溶剂，也是景观……

——辛西娅·林，《尘绘的实际尺寸》，2003[1]

故事缘起于2014年哈佛访学。穿梭在漫天大雪、空气清冽的寒冬剑桥，哈佛设计研究生院的课堂交流节奏紧张而又令人兴奋。伴随着对于建构和热力学等建筑学核心知识的思考，以及对于自身角色与学科危机的不断思辨与讨论，我的视野始终关注着中国城市化进程中需要应对的独特而严峻的挑战。中国很多城市冬季严重的雾霾吸引着我的关注，它几乎无所不在，我们却束手无策。日益困扰中国很多城市的环境问题，以雾霾带来的空气恶化尤为凸显。中国快速的城市化进程，是否必然以恶化的环境为背景与依托？作为建成环境的一部分，这些是否理所应当游离在建筑与城市的专业视野之外？我们是否需要一种立足专业、关注社会生态的更为广阔的视野？

布鲁诺·拉图尔讲"我们生活在实验室"[2]之中，当代社会的变迁使得边界在可见与不可见间变得模糊，在一个更为复杂的系统中，"技术—环境—文化"在当代社会凝结成一个巨大的整体，呈现出复杂的、纠结待解的状态。狭隘的专业划分往往使得我们在面对复杂系统时缺失了思考与介入能力。2015年同济大学建筑与城市规划学院国际暑期学校以"设计应对雾霾——热力学方法论在中国"为主题，第一次将跨越学科、涉及社会生态议题的"雾霾"作为建筑学的研究主题。同时，鉴于热力学引入城市与建筑领域带来的专业范式转变，课题将哈佛、同济等建筑院校推动的热力学建筑前沿研究作为思考这一主题的工具和方法，"热力学已涌现为一个科学的工具，服务于社会规划，甚至是一个新的范式，通过引入熵和不可逆转的时间概念来塑造思想的景观。"[3] 在此，看似与建筑学本体不太相关的议题设置，恰恰是希望逼迫出专业的自我拓展。正如菲利普·汉姆界定的那样："公共空间成为一个品质由空气界定的场所。它的冬暖夏凉的温度与气候、反照率和植被相关，在公共空间创造一个防止污染的程式。空气组织了公共空间的形式、形态与材料。它提升了公共空间作为一种室内无法实现的生理需求，并在此产生社会生活和公众交流。"[4]

"设计应对雾霾"是一个美好的图景，但建筑设计显然无法独自承担雾霾的挑战，对于根本性的源头问题甚至有些无力。雾霾是中国特定发展阶段下的社会进程、经济生产、城市模式、能源利用的综合副产品，显然关乎基于社会政治与经济结构的应对，从一个热力学系统的角度思考，从能量流动的总体循环思考，雾霾又与城市与建筑为载体的实体环境模式与能源模式密切相关。考虑到"建筑与自然"这一亘古的学科关注，我们也有

---

1. http://cynthiaartist.com/drawings-of-dust/, Cynthia Lin, Actual size of dust drawing, 2003.
2. Bruno Latour, Steve Woolgar. Laboratory Life: The Construction of Scientific[M]. Princeton University Press, 1986.
3. William W. Braham, Daniel Willis. Architecture and Energy: Performance and Style[M]. London & New York: Routledge, 2013: 3.
4. http://www.philipperahm.com/data/projects/publicair/index.html, Philippe Rahm. PUBLIC AIR. 2009.

理由相信，在建筑与城市的领域，一定有需要我们重新思考的范式与路径，对于当前的习以为常的建筑与城市模式提出质疑和提升。根据热力学第二定律的观点：城市是一个"开放的非平衡的热力学系统"，那么，我们有理由设问：面对雾霾这样的城市衍生物，主导着建筑与城市设计的建筑师们，是否有能力和方法提出应对之策，同时通过"知识—话语—范式"的讨论拓展学科边界，我称之为"设计应对雾霾"！（图1）

1. 空气—建筑—城市

阿历杭德罗·德拉·霍塔讲"建筑就是空气，空气就是建筑"，[5] 现代主义早期的技术主义大师们敏感观察到建筑物质性躯壳背后的"空气"是一个重要的联系媒介。现代主义建筑肇始，对于实体建构与能量流动、空气动力的同步关注，是现代建筑考古意义上的珍贵遗产。通过建筑实体之外空气的关注，建筑不再是一个与外部隔绝的物体，而在建筑空间、集群、城市与外部的环境、气候、以及在地的文化之间，建立起一种长久的、值得不断探究的深刻关联。建筑成为自然系统的一个动态的、重要的媒介，从而建筑有了"在地"的基础，并与具有地方性特征的气候、环境、材料文化、生活方式相关联。

柯布西耶在南美之行的"精确性：建筑与城市规划状态报告"一书中描写了飞机掠过南美亚马孙森林的场景："太阳蹦了出来，它逐渐改变了周围的气氛，不同密度的气流相互穿越，有些甚至迎面相撞。小团云彩之间的和平秩序被另一种无法抵抗的力量所统治，它们相会、融合、重组。"[6] 对于地域的空气与气氛的相关描述，超越了城市与建筑的领地。而柯布西耶在印度昌迪加尔、阿巴拉巴德的公共建筑与住宅设计实践，塑造了热带地区的独特的遮阳与虚空空间，展示了无所不在的空气感。在哈佛的卡朋特视觉艺术中心，自由架空的底层、穿越的坡道、随太阳的高度角变换的遮阳，均展现了现代建筑对于"空气流动"为代表的环境要素的积极互动与塑造。

因而，当我们把建筑与城市当做一个"建成环境"时，我们已难以将实体的物质化城市与建筑，与城市所在的环境、自然以及社会严格区分。在现代建筑的建成环境议程中，隔离还是融入，是一个不断被追问的话题。瑞纳·班纳姆对于气泡小屋的设计（图2），巴克明斯特·富勒对于曼哈顿穹顶的乌托邦构想，乃至伴随宇航发展带来的全控制的人工环境的可能性，展现出伴随技术发展，在建成环境领域充满乐观主义的"可控环境"野心。而另一方面，面对唯一未被自然化的气候，面对今日全球性的气候变化与环境挑战，除了技术的应对，我们需要从一个超越现代建筑与城市的视野来思考问题，毕竟现代只有百年的时间，而建筑与城市的存在则悠远的多。在中国久远的城市与建筑之中，一种与自然环境充分互动的建造模式、被动式的环境策略、因地制宜的建筑材料、以及跨越广阔地域与气候环境的建筑智慧，是否在当代城市与建筑中依然有效？（图3）

对于现代城市化伴生的雾霾，从社会、政治、经济、健康等各个层面的讨论和争议固然不可或缺，而从建筑、城市与环境层面去展开，设计所做的既可以是批评与批判，更多应该是介入与应对，这是设计应对雾霾的基本立足点。

2. 热力学建筑方法论

热力学是关注能量流动的科学，热力学原理是一个传统物质世界的法则。将热力学定律应用于建筑领域，"空气"便成为空间组织的主角：建筑可理解为一种物质的组织，并由这种组织带来

---

5. 勒·柯布西耶. 精确性：建筑与城市规划状态报告[M]. 北京：中国建筑工业出版社, 2009:5-6.

6. Iñaki Abalos. Interior. Biennale Architettura 2014, 2014.

图2. 气泡小屋,瑞纳·班纳姆

图 3: 阳产土楼民居集群

"能量流动"的秩序，同时平衡与维持"物质的形式"。热力学原理所建构的能量流动原理，成为多领域跨学科的知识，正如伊纳吉·阿巴罗斯教授所言，"通过知识夺回权威",[7] 通过对于热力学知识的汲取和扩展，建筑师可以扩展视野中的社会与生态图景，把握一个应对雾霾环境议题的工具和方法论。

热力学第二定律界定了一个"远离平衡的开放系统"，建筑与城市就是一个巨大的开放系统。物理学家安德鲁·贝让在"自然中的设计"中界定了"系统—边界—环境"的关系，环境中的开放系统具有一个输入与输出的边界（图4）。在生物界，作为热力学子集的新陈代谢是维持各种生命系统运作的物质与能量转换。而建筑实体的建构最终都与一个开放系统中能量流动的系统相关联。"那些非孤立、非线性的热力学系统，如建筑、城市、生态系统以及生命体本身，都旨在完成三件事：以最快的速度、最强的反馈、最大的量来流通并转化能量。"[8]

作为环境系统局部崩溃的雾霾现象，其成因众说纷纭，但不外乎与城市特定发展阶段的产业特征、能源使用与能量循环密切相关，是城市领域无可承受的熵增和建成环境系统混乱的表征。城市过量的化石燃料消耗、城市模式对于汽车的高度依赖、城市设计架构中热岛效应与风道缺失、采暖空调的单一模式以及隔离式建筑排放的累积，当然还有特定季节气候环境调节的不利，基本被认为是中国雾霾的元凶。热力学的思考，显示建筑与城市系统很难从雾霾的应对中独善其外。当我们将建筑当作一个非隔离系统来创造，显示我们不会满足于密闭门窗和室内的空气净化装置，在这些防御性手段之外，城市与建筑作为这一复杂挑战的一个重要环节，显然有太多值得思考和反思的方法和范式问题。

柯布西耶用"回转法则"解释城市与建筑设计上的种种提议，"我慢慢理解了在人类日常生活中遭遇的境地：他们陷入死胡同，忽然又冒出一个绝佳的办法解决所有的困难。"[9] 热力学的思考可否导出一个应对雾霾挑战的"回转法则"？对于建筑中空气的关注，恰恰与热力学的引入息息相关。

3.现代建筑检视

联系到现代建筑对于"建成环境"的忽视，"设计应对雾霾"的思考也是拓展建筑学科边界的契机！伴随对热力学建筑的研究，我们逐渐意识到建筑与城市作为一个建成环境，其具有超出本体设计之外的更多的观察与批评之视角。而新视角反过来促成了新的可能性思考与形式的进展，带来一种新的基于系统思维的方法论之革新。"最终，人们会认识到，真正的建筑热力学，不是简单粗暴的能量分析与优化的抽象概念，而是以另一种更世界性的策略去应对建成环境的根本。"[10] 从乌托邦层面、原理层面、热动力技术层面、以及材料文化层面，随着思考与讨论的深入，我们会发现一度被遗忘的建筑热力学的意义。

长久以来建筑一直试图并已从"原理"的范畴脱离出来，建筑的自主性更多建立在对于"空间—建构—材料"等形式要素的关注，而"原理"沦为美学结合风格背后最弱不禁风的内核。建筑似乎成为一种狭隘的末端形式的创造。"当代建筑日渐关注诸如生态和社会责任等议题，在面对日益强烈的环境与社会遗存矛盾，传统建筑学和

---

7. Kiel Moe. Insulating Modernism-Isolated and Non-isolated Thermodynamics in Architecture [M]. Berlin: Birkhäuser, 2014.
8. 勒·柯布西耶. 精确性：建筑与城市规划状态报告 [M]. 北京：中国建筑工业出版社，2009:4.
9. Kiel Moe. Insulating Modernism Isolated and Non-isolated Thermodynamics in Architecture[M]. Berlin: Birkhäuser, 2014.
10. William W. Braham, Daniel Willis. Architecture and Energy: Performance and Style[M]. London & New York: Routledge, 2013: 3.

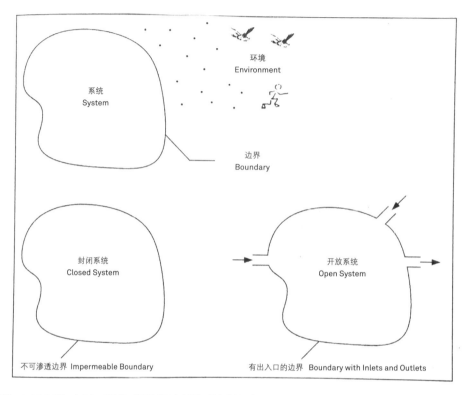

图4. 边界：不可渗透边界与有出入口的边界。假想的零厚度边界将系统与外界隔离开。

理论自主性正受到挑战"。[11] 在面对"环境建构"这一个不得不面对的话题时，建筑师又被商业策划、绿色评估等呆板的标准所绑架，处于一种失语与肤浅应对的状态。

雷纳·班纳姆在从对于现代主义的继承与总结中，关注到建筑环境的重要性，但在"空气穹"与移动汽车建筑等提案中，现代建筑的密闭的观念依然起到主导的作用，它在建立一个与外部隔绝的人工控制环境。建筑与自然的关联在现代建筑进程中实际上变得越来越单一和隔绝。"就自治性而言，建筑形态作为新陈代谢的重要话题在20世纪被简化为形式图解，从热力学的观点来看，建筑学只剩可怜的形式。"[12] 而反思也在持续的进展，迪勒与斯科菲迪奥2002年的"模糊亭"设计展示了一个没有明确边界的融入环境的全新建筑形式，探索了水与气作为一种建筑材料的可能性。匹兹堡市中心更新提案中，周围的空气会对人类的活动做出响应，它可以变化、聚集、移动、重塑以激发新的居住、社会性与社区形式。王子耕针对雾霾空气推出"安铂蒂"计划，不能被当作建筑师的乌托邦的调侃和戏谑，而是提出了建筑师如何介入城市叙事的严肃话题。库哈斯在2014年威尼斯双年展"基本要素"对于建筑基础元素的审视，基尔·莫对于"隔离现代主义"的批评与反思，伊纳吉·阿巴罗斯从热力学观点对于建筑的重新审视，显示了对于隔离的现代主义建筑的检视正当其时。

4.设计应对雾霾

20世纪60年代对于洛杉矶雾霾的研究：提供了程序化分解的方法，去解读复杂的雾霾形成与控制机制(图5)。柴静的"穹顶之下"让人联想到富勒的"曼哈顿穹顶"，前者是对于中国雾霾状况的形象展示，后者则是提出了一个乌托邦在城市尺度上应对和响应环境的可控人造物。托马斯·福尔奇与克里斯·里德对于牡蛎礁湾流的研究图解，展示了建筑与城市尺度上工具边界的日益强大与支撑(图6)。物理学研究也为自然界不可见的流动与能量的展示提供了更多的途径，剪切流场中的开尔文—亥姆霍兹不稳定的演变，通过在两种流动中注入荧光染料使其边界可见，气流翻滚成旋涡，然后相互作用和破裂成湍流(图7)。设计应对雾霾，不仅在方法论的层面，而且在设计与工具的层面，在跨学科的合作层面，均具有操作的可行性。

布鲁诺·拉图预言了科学向研究的转变，并提出在一个科学与社会不可分割的时代，科学的问题不再是一个纯粹的科学，而是通过"社会介入"的方式展现出来。我们可以从拉图尔对于抗生素的研究发现这一路径。设计，尤其城市与建筑尺度的关注，为"雾霾"这一议题假设了一座可能的桥梁：在建筑学专业的界定与社会环境议题的应对之间，我们需要桥梁去突破学科的孤岛。通过来自建筑学、环境学、热能工程学、社会学等不同学科研究者的碰撞，设计应对雾霾，可以成为一个思考、批评与积极介入的契机。

以"设计应对雾霾——热力学方法论在中国"为主题的同济大学建筑与城市规划学院国际暑期学校，集中了16位国际化背景的教授、实践建筑师与跨学科的学者们，以及50位来自包括哈佛、东工大、斯图加特、同济、清华等国内外顶尖建筑院校的年轻学子们。在超高密度与强度的讲座、争论、头脑风暴与设计研讨、评选中，提出了基于丰富主题的研究提案。

设计选址在上海陆家嘴金融商务贸易区。自20世纪90年代，上海政府在浦东陆家嘴地区

---

11. Kiel Moe. Insulating Modernism-Isolated and Non-isolated Thermodynamics in Architecture[M].Berlin. Birkhäuser, 2014.
12. Philippe Rahm. Architecture météorologique [M]. Paris: Archibooks, 2009.

设计应对雾霾 DESIGN AGAINST SMOG

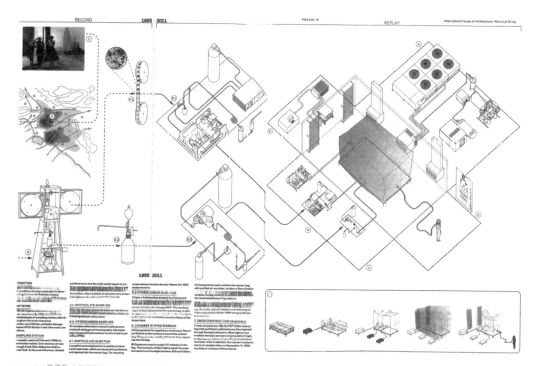

图 5. 对于雾霾形成与控制机制进行程序化分解

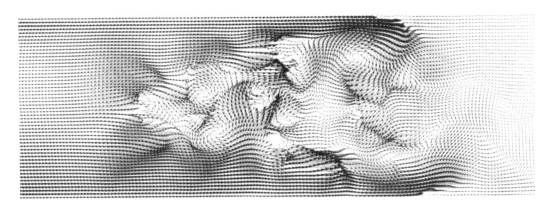

图 6. 托马斯·福尔奇与克里斯·里德对于牡蛎礁湾流的研究图解

进行了雄心勃勃的规划，打造具有国际水准的金融商务中心。如今，陆家嘴 CBD 的占地面积达到 6.8 平方公里，与曼哈顿下城金融区面积相当。陆家嘴贸易金融区在规划阶段就设想了由诸多标志性高层和超高层建筑来构成天际线，其中首个超高层建筑便是 1995 年完工的东方明珠电视塔。另外三座重要的标志性超高层建筑金茂大厦、环球金融中心、上海中心也通过多轮国际竞赛相继建成，如今该区域内的高层超高层建筑总量达到约 100 座，形成非常典型的高层、超高层建筑集群形态。环境学院李卓老师等的研究"陆家嘴地区多尺度流动与扩散的风洞实验研究"为课题展开提供了来自跨学科领域的科学依据。（图 8）

本次项目分为两个阶段：第一阶段针对空气与建筑、城市的关系做原型设计；第二阶段将原型植入具体的城市环境。基地位于上海陆家嘴金融贸易区，有着特殊的城市发展历史。我们的设计目标是针对该区域内的空气污染、气流流动和建筑形态提出概念性的前沿设计。在设计之初，根据初步提案，设计团队被划分成为八个小组，并梳理出八个关键词界定的方向性提示：热力学装置、热舒适实验、生态介入、城市叙事与系统、界面系统、能量与物质化、空气流动形式、集合形态等。

对于从城市、集群、建筑到材料不同尺度的关注，从可见形态到不可见能量的关注，从系统到运作的关注，均得以在研究中得到鼓励。而不同形式的表现与表达，更展现了研究的未来潜力与跨学科启示。

提案一：题为"非'异形'热力学"的提案，基于对于陆家嘴地区的整体通风的环境模拟，发现整体城市建筑集群中存在的死角和恶劣之处，通过一种建筑和热力学复合的组合体，介入其中，极大改善风的流动。设计同时基于系统的视角，塑造一个以步行为主的地区整合策略。

提案二："不可见的基础设施"，道路、步行道、高架环路等基础设施的价值得以重新评估，在热力学意义上，赋予其与建筑垂直方向上的连贯性与一体化考量。从而，通过一种新的基础设施类型的介入，提出较少的汽车排放、加强通风与能量流动的巧妙路径。

提案三："净化：生态介入"通过生态学在建筑层面整合性紧密介入，设计提出了一个新型的可调节界面的概念，通过精巧设计的界面形态、装置与垂直植物的结合，以"绿色"的姿态，展示一个具有空气净化与建筑形式的可能。

提案四："核：空气基础设施"设计具有不凡

设计应对雾霾 DESIGN AGAINST SMOG

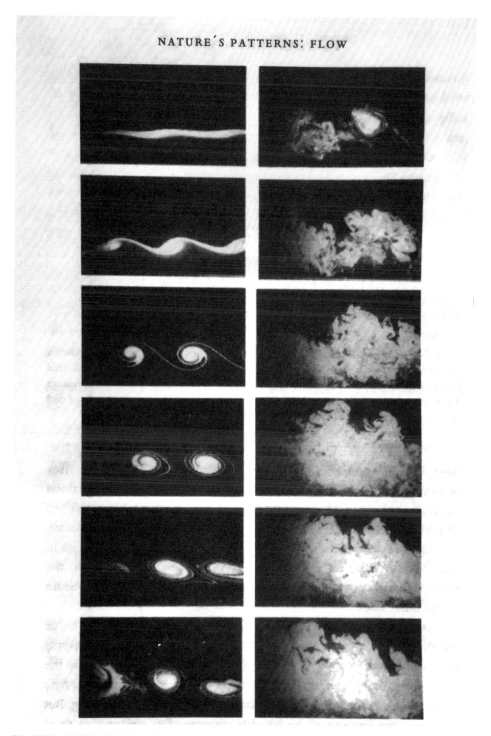

图7. 开尔文—亥姆霍兹不稳定性在剪切流场的演变,通过在两个流的边界注射荧光染料实现可见,波流翻滚汇入漩涡,互动、交融、分解。该序列遵循从顶部到底部,开始向左,进而向右的规律

图8. 陆家嘴地区空气实验

的建筑范式革新,通过对于核心体,尤其是高层与超高层建筑中司空见惯的核心体的重新审视,发现其在热力学意义上的可塑性,从而将其从单纯的管线通路、垂直交通与结构支撑中解放出来,提出了一个整合建构与热力学、具有空气流通与净化作用的新的核心体,力图重构高层密集的CBD区域的空间建构与能量流动。

提案五:"热力学呼吸系统"提出引入"呼吸系统"的概念,将城市看作一个具有新陈代谢功能的有机体,研究雾霾与废弃物排放的减少、流通与排放的热力学原理,提出一个"乌托邦"色彩的超级高层建筑计划,并宣称其在500米高空以上的尺度,成为新鲜空气引入和雾霾驱散的巨大烟囱体。建筑性能的关注使其异于其他。

提案六:提出"藤"的概念,试图通过立体化的基础设施与公共空间的重塑,减少机动车辆的使用和废气排放,关注最为重要的人的尺度层面,创造良好空气品质和公共空间品质。

提案七:"步骤:退出雾霾"提出通过退出平台的概念,关注到"能量—系统—城市"之间的密切关联,借助全球化背景下的信息流动技术,引入一系列的盒子系统、运动系统、气垫系统等构成的装置,并提出借助众筹来实现的策略。

提案八:提出通过"互惠的城市主义"的概念,关注消费与生产作为一个大的城市能量流动的基础规律,重新发掘城市河流的意义,以及其在联通两岸、较少废气排放、促进空气流通、以及创造新型城市模型上的巨大潜力。

5. 结语

中国巨大尺度、规模与速度的城市化,正站在一个转折的十字路口。环境恶化带来的城市模型与范式的不可持续,正面临一个系统熵增的困境。通过热力学的透镜,展示对于建筑中空气研究的一个方法与路径,从而带动建筑学科对于不可回避的可持续议题的关注,并拓展学科的边界和潜力,是本次国际工作营与后续集合研究的最大期望。布鲁诺·拉图讲:"一个学科的边界愈是涉入其他学科,就越有前途!"[13] 希望设计应对雾霾,在建筑学科的边界开拓、建筑学知识生产的新潮流之中,提供一个有意义的前沿探索和跨学科实验。

---

13. Bruno Latour. We Have Never Been Modern [M]. Catherine Porter (trans.) Cambridge: Harvard University Press, 1993: 6.

参考文献

[1] Neeraj Bhatia, Jurgen Mayer H. Arium: Weather & Architecture[M]. Ostfildern: Hatje Cantz, 2010.

[2] R. Buckminster Fuller. Ideas and Integrities: A Spontaneous Autobiographical Disclosure, Lars Muller Publishers, Baden, 2010.

[3] Reyner Banham. Architecture of the Well-Tempered Environment[M].Chicago: University Of Chicago Press,1984.

[4] Ashley Schafer, Amanda Reeser Lawrence. PRAXIS: Journal of Writing and Building, Issue 13: Eco-logics , Praxis Inc., 2011.

[5] Philip Ball. Flow: Nature's Patterns: A Tapestry in Three Parts [M]. Oxford University Press, 2011.

[5] 李麟学.知识·话语·范式：能量与热力学建筑的历史图景与当代前言[J].时代建筑，2015,2.

图片来源

图1：来自网络。

图2：R. Buckminster Fuller. Ideas and Integrities: A Spontaneous Autobiographical Disclosure. Baden: Lars Muller Publishers, 2010: 265.

图3：摄影师：李麟学。

图4：Adrian Bejan, J. Peder Zane. Design in Nature: How the Constructal Law Governs Evolution in Biology, Physics, Technology, and Social Organization. Doubleday, 2012: 39.

图5：Ashley Schafer, Amanda Reeser Lawrence. PRAXIS: Journal of Writing and Building, Issue 13: Eco-logics. Praxis Inc., 2011: 114-115.

图6：Chris Reed (Editor), Nina-Marie Lister (Editor). Projective Ecologies, ACTAR, Harvard Graduate School of Design. 2014: 286-287.

图7：Philip Ball. Flow: Nature's Patterns: A Tapestry in Three Parts. Oxford University Press, 2011: 34.

图8：图片提供：李卓。

Why Air?

Li Linxue

What is air? A giant void, nothingness.
The transparent solvent of the world invisible layers that also are landscape...
——Cynthia Lin, *Actual Size of Dust Drawing*, 2003 [1]

The story began in the winter of 2014, when I was a visiting scholar at Harvard Graduate School of Design (GSD). Shuttling in Cambridge with the snow flakes whirling in the chilly air, I had a demanding yet exciting one-year exchange programme and was engaged in intensive analysis and discussion on the core knowledge of architecture, such as tectonics and thermodynamics of architecture as well as the crisis and role of this discipline. I have always been concerning and thinking about the severe and unique challenges of urbanization in China. The dense smog in Chinese cities in winter attracts my attention in research, it is omnipresent, but any effort seems failing. Air deterioration caused by smog contributes largely to the building environment issues in Chinese cities. If the triumphant advance of urbanization has to sacrifice the environmental degradation? Shall all these fall outside of architectural and urban disciplines? Is a more comprehensive view based on professional knowledge and concerns on social ecology required here?

Bruno Latour once says that "we live in a laboratory". [2] The boundary between visible and invisible has been constantly agiltated in nowdays. In a system more complex like today's world, technology, environment and culture integrate into an enormous entity, demonstrating a complicated and tangled state to be solved. Devides between different disciplines deprive us of the ability to think and intervent whenever confronted with complex systems. Thus, in the International Summer School 2015, themed on "Design Against Smog - Thermodynamic methodology for Chinese Architecture" and hosted by the College of Architecture and Urban Planning of Tongji University, we proposed smog research as the main topic for the first time to promote interdisciplinary research on architecture involving social and ecological issues. This paradigm shifting is plopelled by the tools and methdlogy experimented shaped in the advanced thermodynamic architectural research in Harvard GSD and Tongji. "Thermodynamics has emerged as a scientific tool to serve community planning, or even a new paradigm, by introducing the concept of entropy and irreversibility of time to shape the landscape of thought." [3] It seems less relevant to architecture ontology though, and this agenda aims to break out the barries set among disciplines. As Philippe Rahm defined: "Public space becomes the place where the quality of the air is defined. Its temperature is heated in the winter or cooled in summer in relation to the climate, albedo and vegetation, creating a depollution process in public. Air organizes the form, shape and materials of public space. It updates the meaning of public space as certain physiological need not fulfilled by interior space, where a social life and exchange between the inhabitants could be born." [4]

"Design Against Smog" is a beautiful vision, how ever architecture alone is defficient in facing the challenge brought up by smog, not to mention the source of smog emnission. As a side effects of the social and urban development, which is the certain development process, smog is obviously more related to the larger image of socio-political and economic structure. To consider the cycle of energy flow, through the perspective of thermodynamics, smog is also closely related to energy model and physical environment in which architecture and cities as the carrier. But taking the traditional concern on "architecture and nature" into account, we are confident that there must be certain paradigms and paths for us to rethink and question current mode of architecture and cities. According to the Second Law of thermodynamics: the city is an "open non-equilibrium thermodynamic system". Hence, we can propose that, as the leading characters in builting environment, architects can raise up the solutions for urban derivative such as smog and to expand the boundary of discipline by means of "knowledge-discourse-paradigm". This process, I call "Design Against Smog". (picture 1)

1. Air-Architecture-Urban

Alejandro de la Hota said Architecture is air, and air is architecture. [5] The technicism masters in the early modernism keenly observed that the air in which the material form immersed into is an important medium. The addressing on tectonic and energy flow aerodynamic aspects in the beginning of modernist architecture is a precious legacy in modern architecture. The building is no longer isolated with its external circumstances; instead a profound and strong connection is built among building space, cluster, urban and the external environment, climate and local culture. Architecture becomes a dynamic and significant medium for the natural system and grants itself a down-to-earth" approach with close linkage with local climate, environment, materials, culture and lifestyle.

In his book *Accuracy-Architecture and Urban Planning Status Report*, Le Corbusier described an aircraft flying over the vast forests in South America, The sun jumped out and gradually changed the atmosphere around it. Air in different densities flowed and bumped into each other. Peace and order of the cloud clusters were ruled by another overwhelming force, they began to meet, merge and restructure." [6] The description of the air and atmosphere falls beyond the boundary of city and architecture. In his practice in Chandigarh and Abbas Abad, Le Corbusier developed a unique shading system and void spaces for those public buildings and residentials.In the Carpenter Center for the Visual Arts at Harvard University open space on the ground-floor, crossing ramps, and shading changing with solar elevation angle, all of which well demonstrate the active interactions between modern architecture and environmental elements represented by "air flow".

When we regard architecture and city as a "built environment", it is difficult for us to strictly distinguish the city and building from the environment, nature and society that the city resides in. In the discussion of built environment in modern architecture, to be isloated or integrated is a issue remains undefined. Reyner Banham's bubble dome (picture 2), Buckminster Fuller's utopian idea of a dome over Manhattan, and the possibility of fully-controlled artificial environment with the support of aerospace development indicate the optimistic ambition of a controllable environment in built environment with the support of technology advancement. On the other hand, an approach to think beyond modern architecture and city is required to address the challenge of global climate change and environmental degradation, given that architecture and cities exist much longer than the time of modernism, which comes to the stage just for a century. It is widely seen in Chinese ancient cities and buildings of a harmonious interaction of natural and environment, passive building philosophy, utilization of local building materials, and the wisdom to deal with wide ranges of different climate and geographical circumstances. The question remains now is whether these approaches are still workable in the contemporary urban environment and architecture (picture 3).

For the ubiquitous smog which is concomitant with the modern urbanlization, there have been discussion, controversy and ridicule on the social, political, economic and human health levels. Yet from the perspective of architecture, urban and environment, designers shall take the stand of Design Against Smog to respond to and get more engaged in this issue, in addition to merely criticism.

2. Methodology of Thermodynamic Architecture

Thermodynamics is a science concerned with energy flow, and the principle of thermodynamics is the law of the material world. When the law of thermodynamics is applied to architecture, the "air" becomes the protagonist of space organization. In this way, architecture can be understood as a materialization of matter, which brings a status of order by "energy flow", while balancing and maintaining the "form of the substance". Energy flow principle that constituted by thermodynamic principles has become interdisciplinary. As Inaki Abalos urges, "Grasp authorization through knowledge, [7] the knowledge of thermodynamics provides architects with tools and methodology to deal with the smog by integrating social and ecological concerns.

The second law of thermodynamics defines an open system far from equilibrium. Architecture and cities is one of the huge open systems. According to the "system—boundary—environment" diagram advocated by physicist Adrian Bejan in "Design in Nature ", the open system of environment is delineated by a boundary of input and output (picture 4). In the biological world, metabolism, as a subset of thermodynamics, serves the conversion of matter and energy to maintain the operation of various life systems. The tectonics is eventually associated with the energy flow of an open system.The thermodynamic constitution of non-isolated, non-linear systems—such as buildings, cities, ecosystems, and life itself—will tend to do three things: circulating and transforming the most available energy, at the fastest rate possible, and with the most reinforcing feedbacks. [8]

Smog is a phenomenon of local environmental degradation. Opinions vary as to the causes of smog, but nothing more than the factors related to industry, energy utilization and cycle in a certain stage of urban development, and smog is regarded as a syndrome of unbearable entropy increase and the disorder of built environment in cities Over-consumption of fossil fuels, high dependence on cars, the urban heat island effect, isolated air condition, accumulation of building emissions, and unfavorable seasonal climate are considered to account largely for China's smog. The consideration of thermodynamics indicates that architecture and urban cannot be excluded from the smog challenge. When we take architecture as a non-isolated system, reflections on the existing methodology and paradigm will be required, as architecture and cities serve significantly in coping with the challenge in addition to defensive means, such as indoor air purification devices and closed windows.

Le Corbusier explains architectural and urban design proposals with the "rotation rule", gradually understand the situation in people's daily life: they fall into a dead end and suddenly spring a perfect solution to all the issues. [9] Could thermodynamic approachlead to a "rotation rule" to deal with the smog? It is the introduction of thermodynamics in architecture that draws our attention to air in the architecture.

3. Inspection of Modern Architecture

The neglect of "built environment" in modern architecture and the initiative of Design Against Smog may also mean an opportunity to expand the boundaries of the discipline of architecture. With the study of thermodynamic architecture, we gradually realized that architecture and the city as "building environment" can be viewed and criticised not only in the perspective of design ontology. The new perspectives, in return, have led to potential of thinking and progress, and brought a new approach to the innovation of systematical thinking. "Finally, the actual thermodynamics of building, rather than the violent abstractions of energy analysis and optimization, are understood as fundamental to an alternative, more cosmopolitan in the most literal sense of the term politics for built environments." [10] With profound thinking and discussion on Utopian principle levels, theoretical levels, thermal power technology and material levels, we will discover this long forgotten corner, architectural thermodynamics.

For a long time, the architecture apparently has tried to and also has succeeded in separating itself from the limitation of "principle". The autonomy of architecture lies much in the concerns of form elements such as space-tectonic-materials, leaving the principle being the most fragile kernel in aesthetic style. Architecture seems to be narrowed as the end of creation. Especially in the face of increasingly strong environmental and social remains, the traditional architecture and the autonomy of discipline are being challenged (Sara Whiting). [11] In the face of "environmental construction", architects are kidnapped by the rigid standards of commercial planning and green assessment, so that they are laid back in a state of aphasia and passive reaction.

When inheriting and summarizing modernism, Reyner Banham noticed the importance of built environment, but in his proposals of air dome and removable cars architecture, the concept that a modern architecture should being a closed environment still plays a leading role. This concept aims at establishing an isolated environment controllable by human. The relationship between architecture and nature is actually becoming more and more simplified during the development of modern architecture. In the first case, the metabolically consequential topic of architectural formation was reduced in the twentieth century to diagrams of object shape which in turn, from a thermodynamic perspective, left architecture in abject shape. [12] But reflections are going on. In 2002, Diller & Scofidior displayed a new architecture form, "Blur Building" which has no clear boundary to separate the environment. This building showed the possibility that water and gas could be building materials as well. In the proposal to upgrade the Pittsburgh Center, an idea has been put forward that the surrounding air will respond to human activities. It can change, gather, move, and cast to stimulate new residential, social and community forms. Wang Zigeng proposed his "Empty Plan" to fight against the smog. It was not just architects utopian ridicule but rather, it presented an architect vigorously involved in city development. Koolhaas' study of the basic elements of architecture in 2014 Venice Biennale "Element", Kiel Moe's criticism and reflections on isolated modernism, Inaki Abalos's new reflection on definition of contemporary architecture in the light of thermodynamics, all show that this is high time to have a review on the isolated modernism.

4. Design Against Smog

A Study on the smog in Los Angeles in 1960s provides a method of procedural decomposition to interpret the complex mechanism of how smog forms and how it is controlled (picture 5). Chai Jing's documentary *Under the Dome* reminds people of Fuller's Manhattan Dome. The former demostrated the smog situation in China while the latter depicted an Utopia where human force can respond to the environment in the scale of a city. The diagram of Oyster Reef Gulf steam studied by Tomas Folch and Chris Reed, shows an increasingly strong support of tools in expanding the architectural and urban boundaries (picture 6). Physics also provides more ways for nature to display invisible flow and energy. In the instability evolution of Kelvin Helmholtz in shear flow, the air flows into a vortex, and then interacts and breaks down into a turbulent flow (picture 7). Design Against Smog approves feasible not only methodologically, but also approves operational on designing, function and the interdisciplinary cooperation levels.

Bruno Latour predicted the transition from science to research, and put forward that in an era when science and society are indivisible, science is not purely science, instead, it always presents itself by intervening into the society. We can find this path from Latour's antibiotics research. Design, with particular attention to urban and architecture scale, establishes a possible bridge to the smog issue. Through which, the barrier of architecture discipline and the social environment crisis can finally bridged. Through interdisciplinary cooperation between architecture, environmental science, energy engineering, sociology and other disciplines researchers, design can be a powerful tool for critical thinking and intervention.

With Design Against Smog: Thermodynamic Methodology for Chinese Architecture as the topic of iternational summer school of the College of Architecture and Urban Planning of Tongji University, it brings together 16 professors of architecture, architects and interdisciplinary scholars with international background; As well as 50 students from leading Chinese and international architectural colleges which includes Harvard, Tokyo Institute of Technology, Stuttgart, Tongji and Qinghua, ect. After intensive lectures, debates, brainstorming, discussions, and design evaluation, we present some research proposals based on a number of topics.

The site is located in Lujiazui, the central business district (CBD) of Shanghai. Since the 1990s, carried out an ambitious plan to build a financial trade center with international standards in Pudong area. Today, Lujiazui CBD covers an area of 6.8 square kilometers, equivalent to Manhattan Downtown financial zone. Lujiazui has been envisaged constructing many iconic high-rise and super high-rise buildings to form the skyline on its planning stage, among which the first super high-rise is completed in 1995, i.e. the Oriental Pearl TV Tower. Other three important landmark high-rise buildings, the Jingmao Tower, World Financial Center, and Shanghai Center, and also have been completed after several rounds of international bidding. Now the total amount of high-rise buildings in this region has reached about 100, forming a typical high-rise and super high-rise cluster morphology. The research, Multiscale Flow and Wind Tunnel Experiment Proliferation in Lujiazui, conducted by Li Zhuo's team from the College of Environmental Science and Engineering of Tongji University, provides a scientific basis from interdisciplinary research. (picture 8)

This project is divided into two stages. The first one is to build a prototype design, explaining the relationship between air, architecture and the city. The second stage is to implant the prototype into specific urban environment as shown. Our goal is to propose a conceptual frontier design to deal with the air pollution, the air flow and the architectural form in this area. At the beginning of the design, according to the preliminary proposal, and design teams were divided into eight groups. 8 key words were identified as the focuses: thermodynamic device, thermal comfort experiment, ecological intervention, urban narration and system, interface system, energy and materiality, air flow form, conglomerate form, etc. This study encourages diversity of research interests, ranging from the city, clusters, from architecture to material, from the visible morphologies to the invisible energy, from the system to the operation. The performance and the expressions of the different forms also demonstrated the future potential and interdisciplinary enlightenment of this study.

Proposal 1: DeMONSTERative Thermodynamics. It is based on the overall simulation of ventilation in Lujiazui area. After finding out the blind side of the area, as a complex of architecture and thermodynamics, will be placed here to greatly improve the air flow in this area. At the same time, this proposal, based on the perspective of system design, tends to conduct anintegrated strategy to design an area mainly for pedestrians.

Proposal 2: Focuses on Invisible Infrastructure, so that the function of road, pedestrian road and elevated highway and other infrastructure can be re-evaluated. From the thermodynamic perspective, they are given coherent and integrated considerations from building's verticality.So that by a new type of infrastructure involved, it proposed a path with fewer vehicle emissions, improved ventilation and energy flow.

Proposal 3: It proposed and designed an adjustable interface based on the close integration of ecology in architecture.The combination of interface, device and vertical plants of ingenious design showed the possibility of architecture with air purification presenting in green.

Proposal 4: This design has an ambition of being anextraordinary and innovative architecture paradigm. Through re-examination of the core of highrise buildings, especially that of the high-rise and super high-rise buildings, the fabricability of thermodynamic was discovered. This will liberate the core from the simple pipeline pathway, vertical transportation and support structure. Besides, it brings about a new core integrating construction with thermodynamics, with the function of air circulation and purification. Thus, the design tends to reconfigure the physical constructs and energy flow in CBD area clustered with high-rises.

Proposal 5: It puts forward a concept of the respiratory-system. That is, seeing a city as an organism with metabolic function. On this basis, they will study the thermodynamic principle of reducing and circulating the smog and waste emission. It proposes an Utopian high rise plan. They claim that, at the height of 500 meters above the ground, those high-rise buildings work as huge chimneys which could introduce fresh air and disperse smog. The concerns of performance distinguishes this proposal from the others.

Proposal 6: This concept of Vine attempts to reshape public space and remodel the dimensional infrastructure so as to reduce the use of motor vehicles and waste emissions. It pays attention to the most important scale, i.e. human and tries to improve the quality of air and public space.

Proposal 7: This group pays attention to the close relationship of energy-system-urban(ESC) by proposing the ESC SMOG concept. With the help of information flow technology in the background of globalization, they introduce a series of devices including the box system, motion system and gas ring system. They propose to achieve the goals by crowd funding.

Proposal 8: By putting forward the concept of RECY-Procal Urbanisms, it pays attention to the fundamental laws of consumption and production as the energy flow of a city, and reconfirms the significance of the urban rivers, and its potentials in connecting both banks, reducing emission, promoting air circulation and creating new urban model.

5. Conclusion

The rapid urbanization and the expansion of city scale, puts China at a critical turning point. Urban development paradigm is not sustainable in the background of environmental degradation as the city facing a dilemma of system entropy. Through the lens of thermodynamics, it shows a design approach towards air research in architecture, so as to draw the attention of architecture to the unavoidable issue, sustainability. Research through air also expores the disciplinary boundary and potential architecture. It is the greatest expectation for this summer school and subsequent academic researches. Bruno Latour once said: the more the boundaries of a discipline involved into other disciplines, the more it is promising! It is expected that Design Against Smog be a frontier exploration and interdisciplinary experiment in the trend of architectural knowledge production.[13]

捕捉空气：空气与建筑的几种可能

周渐佳

"你的生命靠它维系，它是脆弱的、技术的、公共的、政治化的，它随时可能崩塌——它正在崩塌，它正在被修复，而你对那些负责修复的人并没有十足的信心。"

——布鲁诺·拉图尔，《空气》[1]

在为期十天的"设计应对雾霾"暑期学校中，被引用频率最高的恐怕是这两张图：一张是笼罩在雾霾之下的陆家嘴（图1），透过不那么透明的空气，只有建筑高耸的轮廓若隐若现；另一张是巴克敏斯特·富勒 1960 年为曼哈顿设计的玻璃穹顶（图2），从东河到哈德逊河之间的四十几个街区都被覆盖在巨大穹顶之下，还有穹顶内数量可观的空气——尽管它们并不可见，但是扮演了提案中最重要的气候调节手段。令人畏惧的尺度使它比富勒其他所有的网格穹顶实验都更有野心，而通过柴静向雾霾宣战的纪录片《穹顶之下》，这幅图像似乎成了表达大城市空气危机的最佳代言。如果说前一张图像描述的是建筑暴露在空气中的徒然，那么后一张则多少带些乌托邦的意味，它是极端的，甚至武断的，穹顶所包裹出来的其实是一座自我封闭的实验室，实验的对象正是需凭借技术手段将之调节到理想状态的物质——空气。这两张图可以代表从 20 世纪 60 年代到现今，建筑学科对空气态度的转变，或许它们也同时说明了两种状态之间的空白：无论后者在现在被怎样评论，甚至被当做技术统治下的极端反例，它都说明了一个尴尬的现实——在过去的五十多年间，建筑学科似乎没有再为建筑与环境之间的关系贡献过如此振聋发聩的范式。

这两张图之间的缺失正是在当下，正是中国讨论"设计应对雾霾"课题的意义，正如空气之于人一样，只有在逐渐失去时才能真正意识到它的存在。最清洁的空气往往是通过不断过滤与净化获得的，这也成了可以用来描述建筑学科在过去数十年内遭遇的隐喻：对学科专业的不断细分使得建筑学一点点摘除自身与环境技术的相关性，"无论管道还是工程顾问，它们代表的是'另一种文化'，它们与建筑的离间如此之深，以至为大多数建筑师所不齿，今天依然如此"。[2] 因此，无论在实践中还是在教学中更为看重的都是"建筑本身"，或者由此而来的一套形式美学法则，而将所有相关的科学拱手相让于"另一种文化"。也许这些是选择陆家嘴作为基地的有趣之处，因为这块基地本身几乎佐证了现代建筑物的"巅峰成就"，挑战天际的高度、差异加剧的地块、高层范式的罗列，在这其中空气问题被诉诸于设备或者一层又一层的表皮，而为了维持恒定舒适的室内环境，这些建筑物俨然已经变成了高能耗、高浪费的机器。

也正是因为这个原因，这个课题比设计一座应对雾霾的机器或者工厂要来得更为困难，因为它所针对的并不是一个特定的对象，而是为已经发挥到极致的现代建筑物提供一个来自当代的补救方法。或许，中期评图时一位导师的发言能

---

1. Latour, Bruno. Air. http://www.bruno-latour.fr/sites/default/files/P-115-AIR-SENSORIUM.pdf.pdf
2. Banham, Reyner. The Architecture of the Well-Tempered Environment[M]. 2nd ed. Australia: SteesenVarming, 2004: 11.

设计应对雾霾 DESIGN AGAINST SMOG

图1. 雾霾下的陆家嘴

图2. 富勒的曼哈顿穹顶

客观地评价这种努力："如果雾霾这个问题已经困扰了全世界那么久,那么就不能把希望寄托于一个只有八天的暑期设计学校……但我们能做的是找到建筑在这场探索中的位置。"这也是邀请来自全世界不同学科背景、不同视角的学生和导师共同参与的原因,我们恰恰需要通过学科的知识来重新找回建筑在新一轮环境危机中的权力。下述的八个提案则多少代表了建筑视野下应对雾霾问题的可能与不足,它们所涉及到的共同点既有可能是求解的基础,也有可能是思维的定式。

第三组的提案将建筑与空气的关系聚焦于表皮,他们提议的是一种轻型的、可变化的附着结构,通过又一层表皮将不理想的环境状态抵御在建筑外。轻型结构上的绿化植栽在进一步过滤空气的同时产生氧气,而结构与原建筑表皮之间的间隙可以被用做另一层阳台等功能的缓冲空间。如果说第三组的策略是从建筑的外部出发,那么第四组则着眼于建筑的内部。名为"核"的提案指的既是建筑的核心筒,也指出了雾霾问题的本质。在观察陆家嘴区域内高层普遍采用的核心筒结构后,该组的提案对历史上高层平面的原型做了充分的研究,并且提出大胆的改造设想:核心筒被当作一部巨大的机器,处理的对象并不是来自于外界的空气,而是将来自于地铁、电梯、中通的被动气流有效地利用起来,在高层拔风的过程中净化空气;同时核心筒本身也不再是封闭而不可见的,展现空气处理全程的同时结合了可供市民使用的公共空间,并且以补偿指标的方式鼓励更多开发商推广这个模式。与之类似的是第五组的提案,他们同样认为现在的空气状况并不是理想的新风来源,所以利用陆家嘴超高层建筑的高度优势从城市上部获取清洁的空气,再通过一系列缠绕在建筑外部的"呼吸系统"向地面输送。这也暗示着作为资本象征的超高层塔楼向服务于公众的基础设施的转换。

显然,无论是采用附加物还是原型转换的手法,上述三份提案更多地是从建筑单体出发。相比之下,其余六组的提案则展现出一些建筑以外的维度。第六组、第二组和第一组都认为原先以车行为主导的陆家嘴模式是空气污染的重要来源之一,所以都将区域内的减少排放、鼓励步行与混合功能作为思考的起点。第六组所提出的藤蔓结构连接起区域内的重要空间节点,在现行的车行交通模式上再叠加了一层为人行和自行车服务的系统,与之结合的是另一层可模数化预制的蓬状结构,以材料中的二氧化钛来分解有害微粒,同时保证藤蔓结构内的热舒适与通风。第二组希望用建筑的方式实现应对雾霾的综合策略,在对区域内的建筑、绿化、公共空间做了分析之后,利用现有元素设计出一套连接了高层的净化走廊系统,一端连接的是鼓励步行的绿地,作为空气进入口;另一端与作为热力烟囱的高层建筑相连,通过空气净化、与季节相应的温度调控与气流运动,在为室内带来舒适环境的同时向外排放洁净的空气。第一组则采用了更大的城市尺度进行设计,通过增加热力学基础设施和混合功能建筑的数量,对陆家嘴的规划模式进行了更加大胆的改造,从工作之城转变为生活之城。上述三组提案都触及到了雾霾产生的原因、基础设施在空气治理中扮演的重要角色以及生活方式的改变,并且尝试将对策演变为一个长期的、动态的也是诸众参与的过程。第八组则提出了一个更广义的设想,认为雾霾产生的原因不是某个区域作用下的结果,而是物质交换的必然结果,所以他们将视野进一步扩大到整个上海,完整演绎了供需生产、运输河道、废物的产生与回收等过程,最终以横跨黄浦江的数座功能复合的桥梁作为设计结果。与以上提案相比,第七组的"能量系统之城"则转向了一种非建筑化的、也更全民化的运动,他们认为传统的空间生产模式已经面临瓶颈,因此需要借助网络与产品的力量,将这种努力进一步轻质化与社会化，为此搭建的网络平台是最重要的战场。

　　这些提案中伴随着对传统建筑、城市模式的反思，对基础设施的依赖，对民众参与的诉求，最重要的是对综合技术手段的依赖，发现这些趋势可能是本次暑期学校课题的最大价值，也同样表明了这门学科面对的局限。无论在学科意义上还是社会意义上，对建筑的定义已经不能局限于墙体或外壳包裹的构筑物，相反，在各种能量、物质、环境的冲击下正在发展出新的形式、审美、组织关系甚至社会体验。事实上，参与其中就是一个从建筑的角度不断捕获空气的过程，是挑战建筑与空气的关系，也是力图撬动现状坚硬的外壳。然而这个过程又是极其困难的，因为所有线索就如同空气一样无处不在，却又充满最细微的影响。

　　在课题制定前不久，炙手可热的BIG建筑事务所和英国设计师托马斯·赫斯维克一同公布了谷歌总部的设计方案（图3），连绵起伏的建筑群被覆盖在透明的玻璃顶下。当然评论家们还在为设计中许诺的私有空间能否真的向公众开放而争论，那轻盈的、透明的、掌控一切的透明玻璃罩却如同富勒穹顶口训的幽灵再现，似是一则来自于过去的未来寓言，但需要思考的是，这则寓言是不是诉说建筑与环境的最好方式？

图片来源

图1：https://www.buzzfeed.com/ryanhatesthis/the-smog-in-shanghai-should-terrify-you?utm_term=.neMJyPOMw#.txM1jLDaX.

图2：http://www.metropolismag.com/July-2008/The-Fuller-Effect/.

图3：http://www.dezeen.com/2015/05/07/linkedin-blocks-big-thomas-heatherwick-proposed-google-hq-headquarters-mountain-view-california/Press, 2011: 34.

图3. BIG和赫斯维克设计的谷歌园区

设计应对雾霾 DESIGN AGAINST SMOG

Catching Air: The Possibilities of Air and Architecture

Zhou Jianjia

You are on life support, it's fragile, it's technical, it's public, it's political, it could break down — it is breaking down — it's being fixed, you are not too confident of those who fix it.

— Bruno Latour, *Air* [1]

During the ten days of the summer school with the subject Design against Smog, there two pictures that have been most frequently referenced: the first one shows smog shrouded the outlines of highrises in Lujiazui (picture 1). The other picture is the glass dome designed by Buckminster Fuller for Manhattan in 1960 (picture 2). From East River to Hudson River, over forty blocks are covered by the huge dome and the impressive amount of air—although it is invisible. However, the air played the important role of climate adjustor in the proposal. The terrifyingly gigantic scale made it more ambitious than other Geodesic Dome experiments proposed by Fuller. Due to the documentary produced by Chai Jing, Under the Dome, the imagery seems to have become the best symbol of air crises in big cities. If the former picture depicts the vainness of buildings exposed in the air, the latter more or less implies Utopia. It is extreme and even arbitrary. The dome actually enwrapped a self-enclosed laboratory. The object of the experiment is the material to be adjusted to its ideal condition with technical means. The two pictures represent the transition of the attitude of architecture towards the air since 1960s to the present. Or perhaps they illustrate the gap between the two conditions. No matter how the latter is criticized now, and even treated as an extreme counter-example of technocracy, it implicates an embarrassing reality — it seems that during the past fifty years, the discipline of architecture has not contributed any enlightening paradigm to the relationship between architecture and the environment.

In the gap between the two pictures lies the significance of projects like Design against Smog. Air is important to human beings. But they wouldn't realize its existence until they began to lose it. The cleanest air is often attained through constant filtration and purification. It has become a metaphor of the situation of architecture in the latest ten years: the continuous subdivision of disciplines has entailed the gradual extirpation of the discipline of architecture of its relevance to environmental technology. "Tunneling and engineering advisors represent 'another culture'. Their alienation from architecture is so profound that most architects hold them in contempt. The situation remains the same today." [2] Therefore, in practice or education, "the architecture itself" and the consequent formal aesthetic principles have been over addressed. All the other related disciplines have been given up to "another culture". The choice of Lujiazui as the base of the project is thus interesting, as the area itself almost exemplifies the "peak achievement" of modern architecture. In the structure constituted by skyscrapers, the land plots in which differences are exacerbated, the spread out of high-rise paradigms, the issue of air is solved with equipments or layers of skins. To maintain a constant and comfortable interior environment, these buildings have indeed become highly energy-consuming and wasting machines.

Due to the same reason, the project is more difficult than the design of a machine or a factory that deals with the issue of smog. It doesn't target at a particular object. On the contrary, it provides a contemporary remedy for the modern architecture that has developed to the extreme. Perhaps the comment of a supervisor during the interim review could objectively evaluate such an effort: "As the issue of smog has troubled the world for such a long time, It is unwise to place too much hope in an eight-day summer school for design… What we can do is to find the position of architecture in the exploration." That's the reason why students and supervisors with various academic backgrounds and different perspectives were invited to participate. What we need most is to restore the power of architecture in the new round of environmental crises through the knowledge of different disciplines. The common part of the disciplines may be the foundation of solutions or the pattern of thinking.

The proposal of TEAM 3 focuses the relationship between architecture and air on the skin. They proposed a light and mutable attachment structure to prevent the unsatisfactory environmental situation outside the building through another layer of skin. The green plants attached to the light structure will further filter the air and produce oxygen. The gap between the structure and the original skin of the building can be used as another layer of cushion space with the function of the balcony. As the strategy of TEAM 3 starts from the exterior of the building, that of TEAM 4 focuses on the interior of the building. Titled 'Core', the proposal refers to the core tube of the building as well as the nature of the smog issue. The team had observed the core tube structure commonly employed by high-rises in Lujiazui and made a thorough study of the historical paradigms for high-rise planes before they proposed a dynamic modification idea: The core tube functions as a huge machine that not only processes air from the exterior environment, but also the passive airflow that comes from the underground railways and elevators. The air will be purified in the high-rise draught. What's more, the core tube itself is no longer enclosed and hidden. The whole process of air purification is displayed and combined with public space to be utilized by citizens. More developers will be encouraged with compensation policies to promote the mode. The proposal of TEAM 5 is similar to that of TEAM 4. They don't consider the present air condition as the ideal source of fresh air. Therefore, they plan to take advantage of the height of super high-rises in Lujiazui to attain clean air from the top of the city. And carried the fresh air to the ground through a series of 'breathing systems' twining around the exterior of the building. The proposal has also implicated the transition of super towers from symbols of capital to infrastructures that serve the public.

Obviously, despite the difference between attachment solution and paradigm transition, the above three proposals deal with individual buildings. In contrast, the proposals of the other six teams have exhibited some dimensions outside architecture.

TEAM 6, TEAM 2 and TEAM 1 all agree that vehicle-oriented mode of Lujiazui is a significant source of air pollution. Therefore, they all start from emission reduction, encouragement of walking and mixed functions. TEAM 6 proposed to connect important spatial nodes within the area with a vine-like structure and superimpose a layer of system that serves pedestrians and bicycles in addition to the current car-oriented traffic pattern. Another layer of modularizable prefabricated fleabane-like structure resolves harmful particles with titanium dioxide and ensures the heat comfort and ventilation of the interior of the vine structure. TEAM 2 hopes to realize a synthetic strategy confronting the smog. After analyzing the buildings, plants and public space within the area, the team designed a purification corridor system that connected high-rises with existed elements. One end is connected to the green area that encourages walking while the other end connects to high-rises that play the role of thermal chimneys. Through air purification, temperature control corresponding to different seasons and airflow movement, the system provides a comfortable environment for the interior and emits clean air to the outside. On the other hand, TEAM 1 conducted their design on a larger scale. They have proposed a more dynamic renovation of the planning mode of LuJiazui through increasing thermodynamic infrastructure and multi-functional buildings. They aim to convert the city of work to the city of living. The above mentioned three proposals have all caught the cause of smog, the important role played by infrastructure in air treatment and the change of lifestyle and tried to change their strategies into a long and dynamic process participated by the public. TEAM 8 has proposed a more general vision. In their opinion, smog is not the consequence of a certain area, but the inevitable outcome of material exchange. Therefore they have further expanded their vision to include the entire city of Shanghai and deduced the entire process from supply and demand production, transport channels, production and recycling of waste. Finally, they presented several functionally composite bridges across HuangPu River as their design solution. In contrast to the above ones, the proposal of TEAM 7 titled 'City of Energy System' turned to be a non-architectural and more popularized movement. From their point of view, the traditional space production model will soon hit a bottleneck. It requires the influence of Internet and products to alleviate and socialize the effort. The online platform constructed for this purpose will be the most important battlefield.

In these proposals, we can find reflection on traditional architectural and urban patterns, the dependence on infrastructure, the appeal to civilian participation and most importantly, the reliance on integrated technical means. Perhaps the discovery of these tendencies is the greatest value of the summer school project. It also proves the limits of the discipline. Academically or socially speaking, the definition of architecture should no longer be limited to structures defined by walls or hulls. On the contrary, the impact of various energies, materials and environments is developing new forms, new aesthetics, new organizational relationship and social experience. In fact, participating in the process means continuously capturing air from the perspective of architecture, challenging the relationship between architecture and air and prying up the hard shell of the present situation. However, the process is rather difficult because all the evidences are ubiquitous as air and full of subtle influence.

Shortly before the research subject was decided on, the prestigious architectural firm BIG and British designer Thomas Heather wick had jointly publicized the design plan of Google headquarters (picture 2). The architectural complex is covered by a transparent glass ceiling. Of course, critics are still debating if the private space promised by the design can be opened to the public. However, the light transparent and dominating glass shade looks like the phantom of Fuller's dome plan. It is a future fable coming from the past. Is it the best way to tell the relationship between architecture and the environment? Perhaps it's the most thought provoking problem.

《北京蓝》的影像叙事

王子耕

《北京蓝》(图1—图10)是《美丽家园》三部曲的其中一部,影片以北京的空气污染为背景,构建了民众通过网络支付平台购买新鲜空气的未来,通过对一个普通北京家庭的日常性记录,描述一个在全球化背景下无法自我供给的生态,探讨了资本主义活动在新的交易形式下的展现,试图表达对数据时代的政治热望的怀疑,以及资本主义地理学景观构架下人的困境。

影片中来自美国的新鲜空气是一种产品,由空气公司安珀蒂(图11)通过波音公司研发,运输时间缓慢,随大气漂流的飞行器运输,无法准确估算送达时间,客户需要在网络平台支付宝进行购买交易。三者看似无关,实则是一种利用特定网络交易进行的合谋。

支付宝是目前世界上最大的电子商务支付平台,每天的交易量超过500万美金。在交易的过程中,以支付宝为代表的交易平台实际上可以利用交易时间差汇集大量的现金流沉淀,这种网络时代的支付形式使支付宝成为一种不需要任何实体和物理空间需求的"云银行"(图12)。这种隐形的基础设施不仅极大地挑战了现有的物理空间格局,同时也创造着支配空间的新规则。利用遍布海底的光纤通讯网络(图13,图14),货币的转移和执行速度以毫秒为单位,与此同时,货物的转移却需要受到距离和物流技术的限制,这两者之间的时间差创造了新的资本机会。按资本最大化的逻辑,为了最大化网络交易平台资金池的总量,我设想了四个原则和与之相对应的策略,分别是:

1. 交易而非只是物品/运输什么不重要,重要的是运输本身(传递空盒子也是有意义的)。

2. 尽可能多的交易/创造需求,去商品性,增加文化暗示(空气不是空气),施加缺乏的恐惧。

3. 交易必须是持续性的/交易物应是消耗品。

4. 尽可能慢的运输/运输时间意味着资金滞留时间,可利用天气等不可抗因素拖延运输时间。

这四个理性的商业策略是整个故事构架和批判的核心,在影片中以马云日记的形式出现。而交易的形成,正是由于全球化时期展现出的新自由主义逻辑和地缘的不均衡(图15)。而如今这种结构性的紧张与矛盾状态被修饰和隐藏在很多不可见的维度里。

全球化实质上是资本主义与生俱来的趋向,是一个寻找他者的游戏。资本主义系统是一个自我溢出性的系统,其自身的稳定建立在向外部的溢出。这种溢出效应使资本主义国家必须不断进行地理扩张和商品交换。冷战之后,伴随技术的发展,资本流动的加速和通讯交通成本的降低使离岸生产的规模不断膨胀,生产到消费的链条在空间和时间维度都变得极其隐秘和复杂,当代永无止境的消费主义伴随着资本主义道德的裹挟,似乎成为20世纪80年代后期全球新自由主义浪潮中的唯一选择。然而,马克思在130年前所分析的"原始积累"的所有特征,在当代仍然强有力地存在着。特别是1973年之后所形成的强大的金融化浪潮,完全展现出前所未有的投机性与掠夺性特征,尤其是轮针对水、空气、森林、土地以及其他公用事业的私有化浪潮,以及对自然资源与社会公共资源所进行的大规模剥夺。片中空气的买卖相当于针对呼吸这种基本生存权的交易,

设计应对雾霾 DESIGN AGAINST SMOG

图1—图10. 电影《北京蓝》剧照

图 11. 影片中的空气品牌 E-MPTY 的商业广告

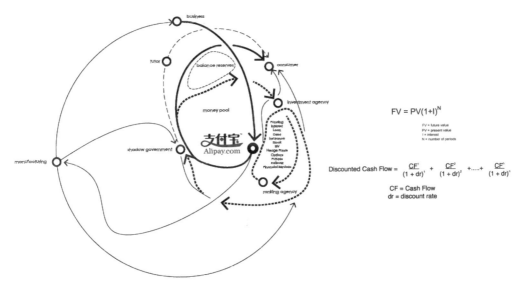

图 12. 网络交易平台的资金流动图解及方程式

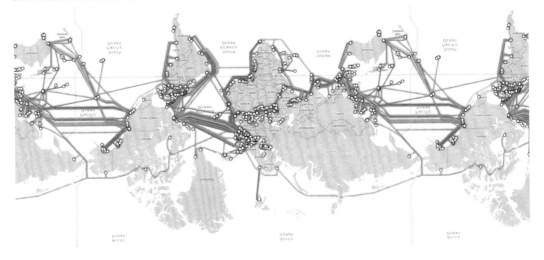

图 13. 海底光纤分布示意图

图 14. 海底光纤断面

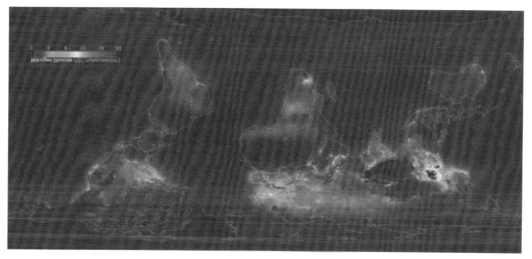

图 15. 空气污染物主要指标二氧化氮含量的全球分布

这些交易真实地存在于 20 世纪许多乌托邦式的社会改良实践造成的悲剧中。

　　列斐伏尔认为，资本主义是通过空间生产而生存下来的，针对空间生产及其逻辑的思考，是人类社会可持续发展宣言的基石。全球化也是一个空间概念，是理解当前城市空间逻辑的基础，而这一观察手段在当前的数字化浪潮里变得更为隐秘和复杂。这部影片所描绘的虚拟未来，正是对现实空间生产资本逻辑的批判和社会图景的警示。

参考文献

[1] 大卫·哈维. 新自由主义简史 [M]. 王钦, 译. 上海：上海译文出版社, 2010：3, 184-187.

[2] 大卫·哈维. 新帝国主义 [M]. 初立忠, 译. 北京：社会科学文献出版社, 2009：28, 54, 128-136.

[3] 詹姆斯·C. 斯科特. 国家的视角：那些试图改善人类状况的项目是如何失败的 [M]. 王晓毅, 译. 胡搏, 校. 北京：社会科学文献出版社, 2012：1-9, 109-112, 117-119.

图片来源：由本文作者提供。

Visual Narration of Beijing Blue

Wang Zigeng

Beijing Blue (picture 1—picture10) is part of the trilogy A Beautiful Country. The film has constituted a future in which civilians buy fresh air through online payment platforms against the background of air pollution in Beijing. Through documentation of the daily life of a common Beijing family, it describes an ecology in a globalized world that can't be self-sufficient. It discusses the performance of capitalist activities under new trading method and tries to express the suspicion of the political ambitions in the age of data as well as the dilemma of human beings in a capitalist geographical landscape frame.

In the film, as a product, fresh air provided by the air company E-MPTY (Picture 11) is transported through a sort of aero-craft developed by Boeing Company which takes much time drifting with airflow whose arrival time can't be precisely estimated. Clients need to purchase the product through the online payment platform Alipay. The three companies seem to have no relationship with each other, but they are actually conspiring through particular online transaction characteristics.

As the present biggest electronic business payment platform in the world, Alipay's daily business volume exceeds five million dollars. During the transaction, payment platforms represented by Alipay can actually accumulate huge cash flow deposits through time difference. The payment form of the age of Internet has made Alipay a kind of cloud bank (Picture 12) that doesn't need any substantial or physical space. (Picture 13, 14) The invisible infrastructure not only greatly challenges present physical spatial structure, but also creates new rules that dominate space. According to the logic of capital maximization, to maximize the total amount of the fund pool of online payment platforms, I have worked out four principles and corresponding strategies:

1. Transaction instead of the object / The object of transportation is not important, what matters is transportation itself (the transportation of empty boxes is still meaningful).

2. As much transaction as possible / Create demand, de-merchandize, add cultural implication (air is not just air any more), exert fear of deficiency.

3. Transaction must be consistent / Objects of the transaction must be consumable supplies.

4. Deliver as slowly as possible / Transport time mean residence time of funds, and can use the weather and other irresistible factors delay transit time.

The four rational business strategies are the core of the structure as well as the criticism of the story. It is expressed with the form of Jack Ma's diaries. The formation of transaction is caused by neo-liberalist logics and geo-unbalance that emerge in the age of globalization. (Picture 15) Today the structural tension and conflicts have been decorated and hidden behind a lot of invisible dimensions.

The essence of globalization is the intrinsic tendency of capitalism. It is a game to search for the other. Capitalism is a self-overflowing system. The stability of the system itself depends on its overflowing towards the exterior. The overflowing effect forces capitalist countries to expand and transact constantly. After the Cold War, with the development of technologies, the acceleration of capital flow and the decrease in communication and transportation cost has led to the continuous expansion of off-bank production scale. The chain from production to consumption has become extremely concealed and complicated. Accompanied by the coercing of capitalist morality, the endless consumerism of the contemporary world has almost become the only choice in the global neo-liberalist trends since late 1980s. However, the characteristics of primitive accumulation' analyzed by Marx 130 years ago still exist forcefully in the contemporary world. The forceful trend of financialization formed after 1973 has exhibited unprecedented speculative and exploitative characteristics, especially in the privatization of water, air, forests, land and other public resources and large-scale exploitation of natural resources and social public resources. The business of air described in the film is equivalent to the trading of the right of breath, the basic right of survival. This sort of trade did exist in tragedies caused by utopian social improvement practices in the twentieth century.

  Henri Lefebvre thought capitalism had survived through spatial production. The thinking of the production of space and its logics is the foundation of the declaration of the sustainable development of human beings. As a spatial concept, globalization is the foundation of understanding the present urban spatial logics. The means of observation has become more concealed and complex in the wave of digitalization. The virtual future depicted by the film is the criticism of and warning against the present capitalist logic of spatial production and social prospect.

应对慢性灾害：烟尘、废气和雾霾

艾奥里奥索托·汉德拉纳塔

本文致力于重新定位建筑在灾害背景下的角色，研究其如何在"极端"的环境或威胁下"生存"。近期中国城市空气中普遍出现的有害微粒的状况，虽然看似没有造成即时伤害，但是这种威胁慢慢造成持续性的环境污染，而人们却对此毫无知觉——这是一种慢性灾害。

大卫·吉森把这种状态称为"亚自然"，即一系列由于其所谓的"危害属性"而被划归至边缘并受到不公正指摘的自然形式，而这些属性其实是值得商榷的，可能同时既有益又有害。[1] 亚自然可以被视作一种不可理喻的景观形式，在这里，崇高而绝对的控制是无用的。或许，它可以被归类为一种当代的景观形式，由技术进步残留的化身集合而成。与印象派画家风景画中的街灯、霓虹灯牌和五彩斑斓的光影相反，亚自然所在的城市不得不与持续出现的焦虑作斗争。[2]

空气中的污染颗粒肉眼不可见，仅仅存在于建筑师空间观的边缘地带，但它已成为影响城市社会、经济和政治方面的重大问题。参考工业革命早期的经验，很显然目前城市环境中的雾霾现象应该理解为一种反复发生的现象，一种不同历史时期不同文明都见证过的事件。

本文将作为一个平台，从空气的文化角度出发，思考建筑与环境之间的关系。本文开始以时间作为叙事线索，呈现一系列对空气的态度，重点讨论空气的不同特点以及它们是否可改变我们的感知以及对空间的认知。这些视觉分析从第一次工业革命开始，吸烟在当时被当作一种社会地位的标识；接下去将围绕一战时期有毒气体的概念，以及控制机动车尾气作为理解城市规划的一种方式进行讨论；随后本文将分析冷战后期诞生的多种生存保障形式，比如巩固的水泥地堡、毒气面罩和密封包装——大多是基于灾难设想，即瞬间发生的灾害而创设的。

本文将把备受雾霾侵扰的北京视为新"空气领域"加以讨论，并提供假设性的计划。例如在区域尺度上采用渐进式的空气修复策略，以此减缓这种慢性灾害的带来的身体损伤等后果。

工业烟雾和瘴气

第一次工业革命见证了生产方式从手工制造转变为机器制造的进步，这也标志着燃煤已经转变为支柱能源。然而，早在五百年前黑色的天空就不鲜见，爱德华一世发布了禁止过度燃烧煤炭的禁令。在1836年题为《对比》的批判文章中，奥古斯塔斯·普金哀叹随着一批新厂房的崛起及其周围有害环境的出现，英国中部的如画风景遭到了破坏。普金的批判中，最重要的一点基于宗教价值：美德的缺失，这似乎建立在教堂尖塔和塔楼消失的基础上，因为它们最终会被越来越高的烟囱替代。

申克尔相信中部地区是烟雾的诞生地，他在前往该地的旅途日记中，详细地记录下他对大气

---

1. David Gissen. Subnature: Architecture's Other Environments [M]. New York: Princeton Architectural Press, 2009: 22.
2. 同上：71.

设计应对雾霾　DESIGN AGAINST SMOG

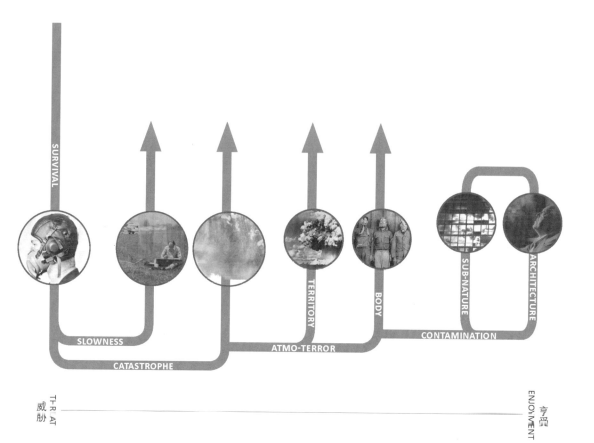

| THREAT 威胁 | | ENJOYMENT 亨受 |

灾难
- 造成巨大的、突然的、破坏性的或痛苦的事件；危机：国家 | 世界似乎正处在危机 | 的方向中
- 戏剧的结局，尤指古典悲剧

慢
- 花费很长的时间来执行的一项具体操作：她是一位阅读速度缓慢的读者 | 大的组织可以慢慢发生变化
- 持续的或花费较长时间的：一个缓慢的过程 | 回家的旅途很漫长
- 不允许快速：慢车道
- 平静而极度乏味：一个缓慢而几近漫无目的的叙述

Air Raid Shelter Queue　防空洞内排队
1939

Great London Smog　伦敦雾霾
1952

Great Beijing Smog　北京雾霾
2013

的观察，以及烟雾的迹象。[3] 你可以推论，维多利亚时期那永远灰蒙蒙的天空和建筑外表的黑色污渍，便是损害人体健康和建筑质量的亚自然形式。从这个时间节点开始，排着废气的建筑形式开始作为一种有缺陷的建筑类型走近日常生活。

尽管如此，约翰·拉斯金最终对废气形式提出了一种不那么理论化的观点。事实上，这位艺术鉴赏家认为占据国内乡村的林林总总的烟囱景观，是美的。[4] 对拉斯金来说，灰色物质的产生与明确的社会阶级划分紧密相连，不同种类的烟雾强调了城市环境类型的差异，即黑色烟雾在工业地区更明显，而权贵阶层则居住在远离黑烟的地方。

大约19世纪前后，在霍乱时期，城市中产生的毒气成为一种污染形式，并导致城市设计发生了一些重要变化。1848年，英国改革家贺科特·嘉文在伦敦东部贝斯那尔格林社区建立了首个测绘毒雾的方法——这是空气污染第一次被测绘。巴黎由豪斯曼执行了类似的项目，对巴黎城市街区的测绘，本质上导致引入更宽的城市道路的结果，这些道路名副其实地把城市分隔成碎片。这个重大变化，使得创造"城市走廊"作为空气通道成为建立净化空气基础设施过程中至关重要的观点，人们相信这些设施能够提供卫生的环境。增加空气流动的同时，还带来了地下排水系统的升级以及对巴黎过度拥挤的街道减负等方面的关键进步。

有毒气体

在《空中颤栗》一书中，哲学家彼得·斯劳特戴克重点强调了1915年4月22日发生的一次事件，他认为这一重要事件标志着攻击环境概念的形成。这一天，德军特种部队用有毒气体攻击法国士兵，他们在敌军到来前不久将毒气释放至空气中。用于攻击的气体不是某种提纯的毒气，而是一些极细微的颗粒，通过爆炸扩散到空气中。[5]

战斗不再在标准的身体层面发生，而是史无前例地发生在大气中，最终感染敌人的呼吸系统。从这一时刻起，人类开始通过操纵自然改变自然呼吸系统的正常功能。对空气发起的袭击使受害者死亡。让·保罗·萨特曾经评论道："人类向自身发起的攻击是令人绝望的。"而对空气实质的攻击则被认为是对人类赖以生存的体制的剥削利用。有毒物质的介入，突显出20世纪恐怖主义、设计意识和环境手段三者结合的实施标准。

德军使用的毒气类型是基于氰氢酸的改良——原本广泛用于烟熏经济作物的杀虫剂。一开始用于加利福尼亚的柑橘类水果农场，很快它的多种用途被迅速发掘。至1917年，2 000万立方米氢氰酸气体已经被用于控制纺织厂、船只、营房、军队医院、学校、谷物和种子仓的害虫。除了毒害效果外，氰氢酸在环境中不会留下任何痕迹，而且没有气味。德国科学家弗里茨·哈伯又进行了进一步改良，制成所谓的A型齐克隆气体，里面添加了一种可感知的刺激性成分，用以警告使用者有害气体的存在。

对于哈伯来说，这一突破也很重要，因为齐克隆气体结合氢和氮，可用来制作氨肥，这对于农作物的大量生产十分关键。有史以来，第一次将饥荒对人类的威胁消除。但是对于德国的化学战计划来说，这为另一种致命的物质的诞生埋下了种子。20世纪20年代初，一种新的有毒物质，齐克隆B气体，在德国工厂中大量生产，这种气体的滥用后来造成了战场和纳粹集中营中大量人员的消亡。

---

3. David Gissen. Subnature: Architecture's Other Environments [M]. New York: Princeton Architectural Press, 2009: 47.

4. 同上: 48.

5. Peter Sloterdijk. Terror From the Air [M]. Frankfurt: Editions Suhrkamp, 2002: 22.

设计应对雾霾 DESIGN AGAINST SMOG

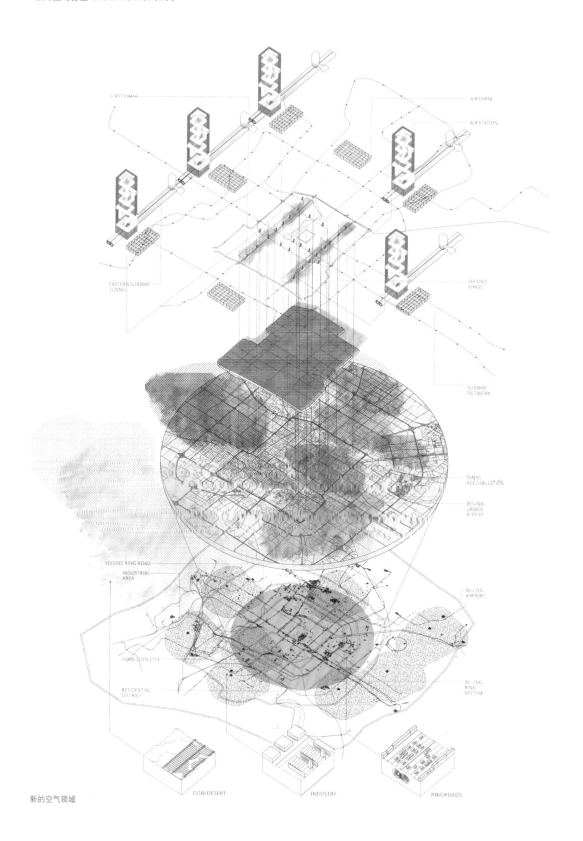

新的空气领域

空气控制

20世纪80年代，廉价汽车的大量生产引发了欧洲和美国公路度量标准的进一步发展。拓宽的道路现在能够容纳更多车辆，而这意味着城市的发展开始向外延推动——工业正是搬迁到这些地方。

因此，20世纪初见证了一次重大转变——污染开始由工业革命时期常见的大烟囱转向城市环境中大量的汽车废气。

基于对更高机动性与日俱增的诉求，加上建立新型高楼的野心，人们开始引入由车流定义的城市街区形态学。城市空间逐渐被二氧化碳构成的生态形式重新定义。1923年，勒·柯布西耶在《高塔之城》中用居民与机动车废气的距离远近描绘了低层和高层住宅的区别。一年之后，柯布在光辉城市的乌托邦提案中，在更大规模的总体规划层面构想了这一原则。方案将居住单元的距离最大化——楼与楼之间的空地用作绿化空间，制造新鲜空气。在创造与地面垂直距离的同时，这些生活单元远离了废气排放发生的层级。与维多利亚时期的工业革命相反，这种分层方式使得接近天空的空气更洁净，而那些地面糟糕的空气，则留给植物去处理。

同年，即1924年，路德维希·希伯赛默基于人们对光线、空气的需要和城市交通的运输需求，通过一项大都会建筑宣言提出，引出了另一个版本的高层城市。[6] 这些街区每个100米宽，600米长，几乎是典型纽约街区的1.5倍，楼体为20层，其中5层商业，其余15层为住宅。街区的巨大面积被提案中60米宽的街道进一步放大，相当于10~12条地面上的机动车道，而互相连接的人行道则被设定在住宅区域的下层。

对于希伯赛默和柯布来说，高层住宅街区有着不同的意义。柯布感兴趣的是一个能够以公园、学校和休闲娱乐空间等形式，创造露天空间的城市。另一方面，赫伯赛摩则开发了一种允许对城市设计采取系统性手段的开发体系，其中的公共基础设施将为居民提供全力支持。

空气妄想症

第一次世界大战之后的1919年，凡尔赛条约签署，标志着德国和同盟国之间达成和解。其中一个主要决议是禁止生产生化武器，包括此前用于改变气候或环境的武器。二十年之后，这些国家深陷第二次世界大战，开始集中发展战争技术。对空气的猜疑和恐惧，在冷战时期再次出现。

由于受到气体袭击的威胁，密封环境或外壳的创造成为建筑创新的一个主要动机。伴随着前两次世界大战的科技进步，这一技术通过充气形式得以表现，HVAC系统的开发使它变成可能。随之诞生的一种设备是1902年发明的电动空调单元，这种设备后来将用于建筑，保证它们在尺度增加之后，拥有相应的舒适气候条件。

阿基佐姆的"无止城市"提案研究了野外环境的概念，在这种环境中，分层的露天楼层能够容纳任何类型的项目，包含了整个生活、工作和休闲娱乐领域。对于构架在周围运转的机械设备，建筑伸缩派称，藏在无尽的地板和吊顶中的管道可以看起来不费吹灰之力地创造环境，能够在封闭的建筑内部提供理想的生存条件。

氰化氢/齐克隆先前的实践中，毒气攻击人类的呼吸系统，接着侵入包括视觉和触觉在内的所有其它消化系统和感觉器官。这种致命气体被进一步改良，能够在更短时间内造成死亡，即从原本的20个小时缩短成20分钟，在此过程中失去生理控制可能造成一种慢性癔病。现在，人们可以辩称，空气调节设备等基础设施系统或呼吸装置的悄然出现——可将原本纯净，现已污染的环

---

6. Peter Sloterdijk. Terror From the Air [M]. Frankfurt: Editions Suhrkamp, 2002: 18.

设计应对雾霾 DESIGN AGAINST SMOG

空气车站模型

境进行替代,这也是抵御此类袭击唯一的办法。

新领空

自中国进入工业革命时期以来,北京就深受俗称PM2.5的空气微粒的危害。这种粉末作为工业废气和市中心机动车尾气的共同产物,被排放到北京的空气中。作为严重雾霾的后果,原本被定义为危险级别的微粒直径标准(5微米)被分类为正常。新的微粒(2.5微米)大约是人类发丝直径的1/30。现在北京面临的是一种缓慢但是稳步发展的灾害,这些微小但致命的微粒,逐渐改变了作为生命之源的环境。

最初引起国际广泛关注是在2008年奥运会期间,最近北京空气污染指数越来越糟糕,已经在可接受的空气质量指数范围之外。政府已经采取措施迅速补救这一问题,比如限制机动车使用,关闭部分北京郊区周边以燃煤为基础的产业。尽管如此,这些措施在治理空气污染方面收效甚微尚待进展。

中国目前每年死于室外空气污染的平均人数已达30万到50万,每年用于治理空气的费用高达数百万美元,北京无法再依靠温和的健康政策和空气税收来解决这一问题。[7] 以现在的破坏率看,这些微粒将最终迫使政府采取城市尺度的物理干涉。即提供一系列空气屏障,作为庇护节点,为市民的健康提供保护。在反乌托邦设想中,雾霾越来越浓重,完全遮蔽了城市景观,城市或许会被迫考虑创设以空气站为形式的干涉——一种临时庇护所,空气灾害的负面效果不断地被减缓。

空气站

洁净的空气越来越成为一种商品,城市环境调节的概念将被引入北京的城市结构中。在创造空气防护系统时,北京将很可能评估其现有的城市基础设施。先前曾得到大力发展的是北京的地下网络系统。这一网络位于城市核心区的地下,目前被大量农民工占据,这些人负担不起地面寓所的房租,只能生活在建于走廊空间中的屋子里,这些地方常被用于种植蘑菇等作物。与此同时,北京持续扩张的地铁网络可以从两方面服务城市:提供更多交通运输节点,以及创造更大的地下集合空间。

利用这两个网络,以空气站的形式从地面对地铁站进行战略性扩张,同时其侧面与地下通道相连,在此共同作用下,创造出一种和雾霾平行存在的生活环境。空气站是一种全新空间类型,由不同气流塑造的环境构成。当大部分外部环境受到雾霾污染的时候,空气站将开采较清洁的空气,移动的地铁列车将这些空气在管道中推动。空气向上移动到人们的避难空间中时已经得到了净化。

另一种折叠管道式的空气腔系统可以从外部得到较脏的空气,进行过滤,然后把净化过的空气重新排到室外。以这种方式,空气站在可同时容纳洁净和肮脏空气系统的结构基础之上建立起来。

结论

纵观建筑史,存在着一种与环境污染密不可分的空间生产传统。从空气的案例中,我们已经看到了我们的态度如何在将它视为一种享受/一种珍贵的呼吸资源/一种威胁/危险的亚自然形式之间动摇。在这些关键时期,有两种建筑思维模式对此作出回应:一种是发展建筑类型,另一种是发展技术去应对特定的挑战。

另一种同时作为实用和防御空间混合发展的建筑形式是疗养院,用于治疗肺结核病人。在

---

7. Ludwig Hilberseimer. Metropolisarchitecture. New York: Columbia University, 2012: 124.

设计应对雾霾 DESIGN AGAINST SMOG

自上而下：空气外流（0 m）；数码商城（-20 m）；储存仓库（-40 m）

应对慢性灾害：烟尘、废气和雾霾 Engaging with Slow Catastrophe: Smoke, Gas, and Smog

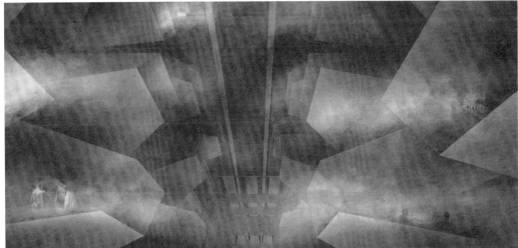

自上而下：日光浴场（+180 m）；浴室（+120 m）；垂直温室（0 m）

Engaging with Slow Catastrophe: Smoke, Gas, and Smog

Erioseto Hendranata

帕米欧肺结核疗养院的设计中，阿尔瓦·阿尔托通过在主体上设置一系列侧翼，同时又能通过家具细节来表达让病人的身体远离污染源的愿望，比如水池轮廓、门把手和躺椅。

不可否认技术发展日新月异，对于中国现阶段的雾霾问题的解答，也需要依靠建筑提案的能力，因为它可以作为一种社会赖以生存的文化工具。只有我们接受它们的存在，让危机和建筑能够共存之时，我们才能真正通过延缓慢性灾害发展的速度，进而与它斗争。

Situated within the investigation on the notion of "survival" under an "extreme" condition or threat, this essay seeks to re-calibrate the position of architecture within the context of catastrophe. One particular phenomenon that has been recurring lately is the appearance of the dangerous airborne particles found within Chinese cities. Instead of instantly appearing perilous, this form of threat occurs slowly to form corrupted environment, which is produced continuously yet experienced unconsciously — a slow catastrophe.

David Gissen argues that these conditions can be understood as Sub-nature: a group of peripheral and unfairly criticized form of nature due to its threatening properties that can actually be argued as both beneficial and detrimental at the same time.¹ Sub-nature might be regarded as an unacceptable form of landscape -- where the idea of the sublime and absolute control seems to be rejected. Perhaps, it can be categorized as a contemporary form of landscape that is assembled by the embodiment of residue of technological processes. Quite the opposite to the scenic scenes portrayed through the street lights, neon signs, and the colored shadows by the Impressionist, the cities where sub-nature inhabits have to deal with the constant presence of anxiety. ²

Invisible to the naked eyes, and often only existing in the periphery of architects' conception of space, airborne contaminated particles have become a major issue affecting the social, economical, and political aspect of urban life. If one refers back to the early Industrial Revolution, it will be apparent that the current phenomena of smog in the urban context should be understood as a recurring phenomena; an event that have been witnessed by different civilizations at different time period.

This writing will serve as a platform for questioning and investigate the relationship between architecture and the polluted environments through the discussions of the cultural aspects of air. It will begin by presenting a series of attitudes towards air in the form of a narrated timeline, highlighting the different characteristics of air that in turns alter our perception and understanding of space. These visual analysis start from the first Industrial Revolution, when smoke was considered a signifier of the social status. This will be followed by the discussions around the notion of toxic air, as a result of the World War I period, and the air controls, as a way that vehicular exhaust contributed to understanding urban planning. It will continue with the later Cold War period, which saw many forms of survival responses, such as reinforced concrete bunkers, gas masks, and a hermetic envelope -- mostly based on the scenario of calamity, i.e. instant catastrophe.

The essay will discuss the new air territory that is produced by the smog by using Beijing as a site for discussion and for speculative project based on a position of a Dystopic scenario, i.e. where incremental strategies of air remedy at the territorial scale will explore the possibility of slowing down the consequences of this slow catastrophe, in which our bodies are slowly being corrupted.

Industrial Smoke & Miasma

The first Industrial Revolution saw the advancement of production methods shifting from hands to machines and this signified a change in the burnt coal being its backbone of energy. However, the black sky was not at all a shocking scene as five hundred years earlier, King Edward I had decreed a ban on excessive coal-burning. In his critical essay of 1836 titled Contrasts, Augustus Pugin lamented on the less picturesque landscape of English Midlands due to the rise of the new array of industrial complex with the production of the harmful environment around them. Most significant to Pugin's criticism was based on a religious value: the absence of good moral, which was likely to be based on the physical disappearance of the church spires and towers, that were eventually replaced by the increasingly taller smoke stacks.

During his trip to the Midlands, the region he believed to be the birthplace of smoke, Schinkel extensively recorded in his diary the observations that he made on the atmosphere and the trace the smoke produced.³ One could assume the ever presence of the grey sky or the black smears on building facades as two forms of sub-nature that degrade both health and buildings during the Victorian Era. From this point onwards, the architectural form of exhausts were closely associated as a defective architectural typology.

Nevertheless, a less doctrinal perspective on the exhaust form was eventually adapted by John

Ruskin. Indeed, the art connoisseur considered the landscapes where the presence of the chimney figures, dominating the countryside domestic settings, to be beautiful.[4] For Ruskin, the presence and production of the grey matter were tied to the reference to the distinct social class divisions, where the different type of smokes emphasized on the difference on the type of urban settings, i.e. the black smoke was more apparent in the industrial areas, while those from higher class lived in places farther away from it.

Around the 1800s, during the Cholera Epidemic, the production of miasmatic air in the city became a form of contamination that brought some major changes in urban design. In 1848, Hector Gavin, an English reformer, established the first method of mapping the disease mist in the neighborhood of Bethnal Green, and East London — the first time air contamination were mapped. Similar initiative was executed by Haussmann in Paris with the mapping of Paris urban blocks that inherently ledding to the introduction of the wider urban corridors that literally cut the city into fragments. With this major change, the idea of creating air tunnels in the form of urban corridors was critical to the establishment of infrastructure for the good air that was believed to provide hygienic condition. Increasing urban air flow also resulted in the key upgrade to the underground sewage system as well as to the process of alleviating the overcrowded Parisian streets.

Toxic Air

In his book Terror from the Air, the philosopher Peter Sloterdijk underlined the importance of an event, occurred on 22 April 1915, as an important incident that marked the conception of attacking on the environment. On this date, the special German force unit attacked the French soldiers using a noxious cloud of gas that were released into the air a few moments prior the arrival of their opponents. This was hardly a pure concentration of gas, instead an extremely fine particles that were spread out through the explosion in the air.[5]

No longer did the combat happen on the normative bodily sense, but, first, it took place on the atmosphere that eventually infected the respiratory system of the adversaries. This was a moment when environment was engineered to alter the norm of natural breathing system. The assault on the air forced the victims into the state of demise. Jean-Paul Sartre once remarked that "desperation is man's attack against himself", and such thing as attack in the substance of air is considered an exploitation of one's life-sustaining habitus. The insertion of a toxic substance highlighted the 20th Century operative criteria of the combination of terrorism, design consciousness, and environmental approach.

The type of gas used by the German was based on a development of Hydrocyanic Acid — widely employed as pesticides for the fumigation of valuable crops. Initially utilised in citrus fruit farms in California, its versatile use was quickly pointed out and by 1917, and twenty million cubic meters of hydrogen cyanide gas had been used to control pest in mills, ships, barracks, military hospitals, schools, grain, and seed silos. Despite its toxic effect, the Hydrocyanic Acid gas left no trace in the environment and were scentless. Fritz Haber, a German scientist perfected this into what was known as Zyklon A, where a perceptible irritant component to alert users to the substance's presence was added.

For Haber, this breakthrough was also important as Zyklon gas synthesised Ammonia for fertiliser from both Hydrogen and Nitrogen, which was the key to the production of more crops. For once, the threat of famine was banished, but for the German chemical warfare program, this was a seed for yet another form of malignant substance. In the early 1920s, a new type of toxic substance, Zyklon B gas, were produced at a high volume in the German factories, and its misuse later led to the deadly form of human extermination both on the warfield and in the Nazi concentration camps.

Air Control

During 1880s, the mass production of affordable cars had led to further evolution of the standard of road metrics in Europe and America. The resized streets were now able to accommodate more vehicles, and this meant that the growth of cities was pushed towards the outskirts — where industries relocated to.

Therefore, the early 20th Century saw the significant shifts from the smokestacks present during the Industrial Revolution to the multiplied number of car exhausts within the urban setting.

The increasing demand on higher mobility combined with the ambition to build new form of high-rise begun to introduce urban block morphology which was defined by vehicular flows. Urban space was gradually redefined by the form of ecology made of the Carbon Dioxide. In 1923, Le Corbusier illustrated in A City of Towers about the distinction between the low rise and the high-rise dwellings in terms of its inhabitant's vicinity to automobile pollution. A year later, Corbusier drafted the principle in a larger master plan in his utopian proposal for Ville Radieuse. The scheme maximizes the distance between the dwelling units — leaving the ground as the green space where fresh air is produced. In creating vertical distance off the ground, the living units get farther from the level of exhausts, where emissions were produced. Contrary to the Victorian Industrial Revolution, this stratification accommodate cleaner air closer to the sky, while leaving the vegetations to take care of the bad air.

At the same year, in 1924, Ludwig Hilberseimer proposed a version of high-rise cities by putting forward a manifesto for a metropolis architecture which was based on the needs for the light, air supply, and the transportation requirements in the city.[6] The blocks, each measured 100m x 600m, almost 1.5 times the size of typical New York blocks, consisting of twenty stories — five commercial and fifteen residential. The vastness of the grid is amplified by the proposed 60m wide streets — equal to ten to twelve lanes, vehicular streets on the ground — while interconnected pedestrians walkway are placed on the lower level of the apartment portions.

For both Hilberseimer and Corbusier, the high-rise dwelling blocks meant different things. Corbusier was interested in a city that could afford the creation of open-air space in the form of parks, schools, and leisure spaces. On the other hand, Hilberseimer was developing systems to allow for systematic approach to urban design, where public infrastructure would support its inhabitants.

Air Paranoia

After the end of the World War I, the Versailles Treaty was signed in 1919. The agreement produced terms for reconciliation between Germany and the Allies. One of the main resolutions was a ban on production of warfare agents including those previously used to alter the climate or environment. Two decades later, the countries involved heavily in the WWII concentrated their efforts on advancing warfare technological developments, and once again, the paranoia of air was reinstated during the Cold War period.

With the threats of attack from the air, the creation of hermetically sealed environment or envelope became one prime motive of architectural invention. Coupled with the advancement in the previous two wars, this technique found its expression in the form of inflatables and made is possible by the development of the HVAC system. One device that enabled this was the electric air condition unit, invented in 1902 that would later empower buildings with provision of climatic condition as their scales go up.

Archizoom's No-Stop City proposal investigates the notion of field condition where stratified open plan floors were able to house any types of programs, encapsulating the entire sphere of live, work, and leisure. Structured around the performance of mechanical devices, Archizoom argued that the production of environment that was seemingly effortless with all the ductworks, hidden within the infinite ground and ceiling planes, was able to provide an ideal environment inside the hermetically sealed building.

In the previous Zyklon scenario, the gas attacked one's respiratory system, and practically all other system of digestion and senses, including sight and touch. The lethal gas was developed further to cause death in shorter time, i.e. from 20 hours to 20 minutes, during which loss of physical control could cause slow form of hysteria. One can argue that, now, the vulnerability of this form of attack can only be tempered by the inconspicuous presence of the infrastructural system of condition devices or survival apparatus — the gas masks and the oxygen breathing device — to substitute the once pure environment that now had been corrupted.

New Air Territory

Since China entered its Industrial Revolution phase, Beijing has been suffering from the appearance of the microscopic particles commonly known as PM 2.5. These dust came into Beijing's air mixture as a combinatory product of industrial air waste and vehicular exhaust within the heart of the city. As a result of the intense smog, the previous standard of particle size that was classified dangerous (5 microns) had been shifted to be categorised as normal. The new microscopic particles (2.5 microns) are approximately measured 1/30th of human hair diameter. What Beijing sees, now, is a slow form but steady catastrophe, where the small, but deadly particles, gradually modify the environment of the source of life.

First brought into international attention during the 2008 Olympic games, and recent readings of air pollution in Beijing have significantly been getting worse by falling outside the acceptable air quality Index. The government have taken measures to quickly remedy this problem by putting a limit on vehicle usage and partially closing the coal-based industries around Beijing's outskirt. Nevertheless, these have shown little or no progress in dealing with these air-borne pollutant.

With the current average of between 300,000 to 500,000 people dying due to air outdoor pollution in China, and millions of Dollars spent annually on cleaning up its air, Beijing cannot afford to rely on its soft health policies and its air tax anymore.[7] In its current destructive rate, the particles will eventually force the government to, undertake physical interventions at urban scale, i.e. provisions of a series of air shelters as nodes of refuge that have to provide both protection and recovery of the citizen's health. In the Dystopic scenario where the intensity of the smog thickens and the cityscape disappearing, the city may be forced to consider creating physical interventions in the form of air stations -- a form of temporary relief where the negative effect of air catastrophe is perpetually slowed down.

Air Stations

As clean air is increasingly becoming a commodity, the concept of urban condition should be introduced within Beijing's urban fabric. In creating an air defense system, the city will be most likely to evaluate its existing urban infrastructure. One that was previously developed is Beijing's underground network. Existing beneath the surface of its city centre, the network is currently occupied by many factory workers, who cannot afford paying for rents for ground property, living in rooms built around the corridor space where food supplies, such as mushroom, can be grown. Meanwhile, Beijing's continuous expansion of its subway network can serve the city in twofold: more transit nodes and creating possibilities for bigger underground gathering space.

Taking the advantage of the two networks, strategic extension of the subway stations overground in the form of air station, and laterally connect with the underground tunnels can work together to provide an environment to create parallel existence of both life and the smog. The air station is a new typology of space that is made of environments shaped by different airflows. In the times when much of the external condition is corrupted by smog, the air station will mine cleaner air pushed across the tunnels by the moving subway trains. The air is cleaned as it travels upwards into the spaces where people find refuge.

Another system of folded tunnel-like air chamber takes in the dirtier air from outside, filters it, and releases the treated air back outside. In this way, air station is created by the structure which accommodates the coexistence of both the clean and dirty air systems.

Conclusion

Throughout the history of architecture exists a lineage of the production of space that is inseparable from the corrupted environment. In the case of air, we have seen how our attitude fluctuates between treating it as an enjoyment — a valuable breathable resource — and as a threat — dangerous form of sub-nature. There are two mindsets of architectural response which have been developed during such critical times: a development of building typology, and a development of technologies to alleviate the specific challenges.

One example of this hybrid development, an architectural form of utilitarian and defensible space, is the sanatorium, which is developed as a space to treat the Tuberculosis patients. In his Paimio Sanatorium, Alvar Aalto was able to spatially configure a series of wings needed to separate the patients while at the same time embodying the desire to enable keeping the patients' bodies away from the contaminants through its furniture details, e.g. the sink profile, door handles, and the reclined chairs.

Without discounting the rapid technological development, the response to the current smog phenomena in China will also need to rely on the ability of architectural proposal that can serve as a cultural tool to enable the survival of the society. Only at the moment when crises and architecture can exist side-by-side — with being accepted their existence by us — then, we will truly be able to engage in slow catastrophe by experiencing it at an even slower pace.

新邻里模型
新城市主义的
热力学

谭峥

邻里与本地化价值

现代化是与城市机动性的发展同步进行的。所有关于20世纪大都会的想象都与高度发达的交通系统有关——从城际轨道到高速公路，从渠化的交通网络到步行天桥，从高速电梯到步行履带。这些技术设施无一不是促进物资、劳动力与资本流动的机器。现代化也是建立在自由流动所依赖的高能耗基础上的。自"邻里单位"这一概念在20世纪初兴起，它就是为了对抗高能耗的机动性并在全球化时代重构"本地"的热力学价值。这一价值不仅在能量领域，也在空间、社会与文化领域发挥着作用。经过数代都市主义学者的探索与更新，邻里模型始终试图创造紧凑高效的社区结构，优化职住平衡与公交服务圈，抑制私人汽车的使用，最终在人居形态层面减少无谓且低效的交通容量与能量消耗。

从广义上看，"邻里"就是历史上的各种社区形式在现代建筑学与规划学中的法定称谓。这种情况下，它不对所指社区的形态、组织与规模做更细致的规定。从狭义来看，产生现代邻里观念的"邻里单位"一词代表了从原理，到方法，再到实践的完整的纵向贯通的城市设计体系。"邻里单位"正式诞生于1913年的芝加哥城市俱乐部的社区设计竞赛。邻里单位在20世纪20年代成为流行于建筑师与规划师中的不成文的原则，并在1929年被规划师克莱伦斯·佩里在规划报告《纽约及周边的区域规划》中作出了更详尽的定义（Johnson，2002）。它最初仅仅是创造一个适于学龄童的步行环境，随后逐渐发展为一种自我完备的生活圈空间模式。

佩里给出了邻里单位所需遵守的六大规则，并绘制了一系列图解以说明邻里单位的典型形态。1931年，佩里对邻里单位的解释被胡佛政府（1929—1933）吸收，在政府的住宅政策中逐渐推广，成为社区规划与地产开发的标准模式，其影响力持续四十年，并在20世纪80年代兴起的新城市主义运动中被再次拾起，成为当代社区设计的重要理论源泉。近十年来，由于全球化的日益深化，由于地缘临近所构成的城市学基础再次遭遇挑战，新城市主义的形式导则的探索渐入瓶颈（Parolek & Parolek，2008）。在这种情况下，需要对新社交模式下的传统物理空间的反应做出系统的历史回顾，尤其需要在近三十年的高速城市化历史中寻找邻里模型的新范式。

泛滥的机动性与新城市主义

从历史视角来看，在二战以后的20世纪中叶，欧美发达国家均一度以现代主义的区划观念来对城市进行功能分区，鼓励工作与居住分离，纵容私人交通工具的泛滥发展，这种以高速公路、高架道路隧桥、快速干道与大街区为主要表现形式的发展方式也在20世纪80、90年代主导了我国的城市基础设施更新，在局部提升大都市交通效率的同时，牺牲了历史中形成的传统小街区，并恶化了城市中行人与非机动车的通行环境。自20世纪80年代开始，美国的城市学家开始反思高速公路蔓延所造成的环境恶化问题，这一时期出现了许多新的针对郊区化问题的规划理论，基于邻里模型的"新城市主义"是其中的代表。 狭

设计应对雾霾 DESIGN AGAINST SMOG

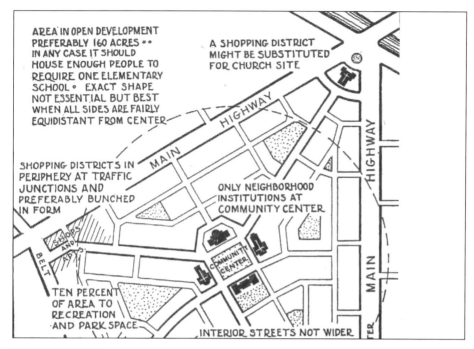

图 1. 佩里的邻里单位模型

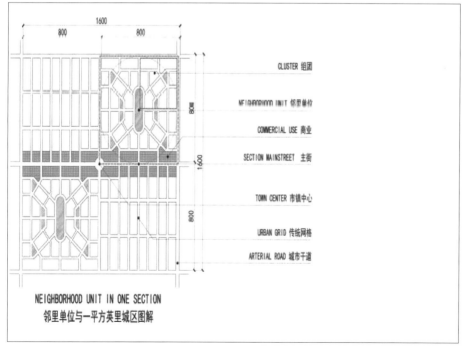

图 2. 邻里单位与典型的美国街区的对比

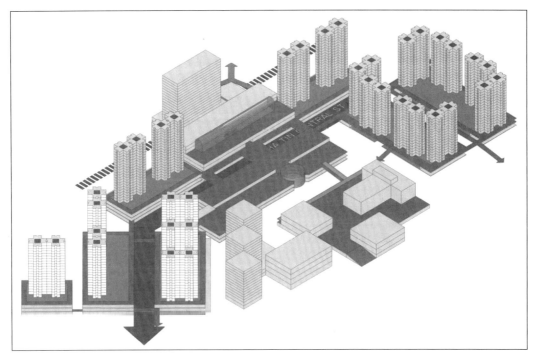

图 3. 香港沙田市镇中心的街区形态

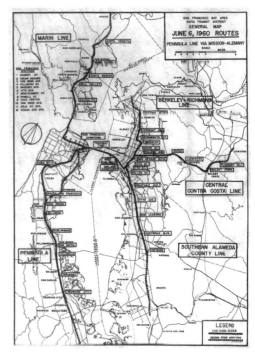

图 4. 1960 年的湾区 BART 系统

义的新城市主义理论主要致力于修补城市扩张过程中丧失的街区尺度与邻里关系，提倡传统（欧洲）的街区形态与本地化生活方式。广义的新城市主义创建以公交系统为主导的区域—市镇—邻里—街区—建筑的一体化形式导则，并对已有的公共空间、道路结构与景观系统进行优化，以改变市民的出行方式。在宏观层面，新城市主义对区域城市的理解与当代经济地理学家的理论吻合，比如布雷纳与杰赛普等人的区域城市理论。在中观层面，新城市主义运用"城市断面样例"等方法进行城市理想密度与类型的推导。在微观层面，新城市主义主张符合文化惯性与形式认知的建筑控制导则。广义的新城市主义是一个综合性的知识体系，而同时期其他一些的城市理论与环境认知理论依然无视托马斯·库恩的后实证主义科学哲学，依然在假设完全脱离对象的研究者，不顾城市理论是一种行动中形成的总结。

尽管有一些批评存在，不可否认的是新城市主义提供了当代城市实践中最高产的策略与模式。新城市主义对美国当代城市设计实践的影响很大，直接指导着许多社区形式规范的编制。新城市主义面对的是业已完成城市化（郊区化）的区域，通过最少的环境干扰与较高的投资效率来加密城市，降低能量消耗，减少汽车出行，改善公共空间。目前我国的城市化进程已经从高速发展进入到中速发展时期，从"增量规划"阶段进入到"存量规划"阶段，许多大都会地区已经进入土地供应的瓶颈，原有的粗放式开发模式已经不可持续。新城市主义的初衷即是对小汽车泛滥的已建成的市区（郊区）进行重构，对我国目前的规划现状有极大的参考意义。尤其需要注意的是，很多以"新市镇"为名的新城建设打着"新城市主义"的旗号，狭隘地将新城市主义理解为某种建筑风貌，却在具体规划措施上与新城市主义的原则有很大出入。

新城市主义与公交都市的成功案例

新城市主义以及公交主导的发展区的概念最初是针对美国的城市规划弊病，后来逐渐影响到新加坡、中国香港等亚洲高密度都市。目前普遍在规划学界公认的符合公交都市标准的北美及亚洲城市包括美国华盛顿特区、西雅图、旧金山、加拿大卡尔加里、新加坡、中国香港等。以下试举两例：

1. 以旧金山为例，旧金山市政交通管理局（SFMTA）控制了整个旧金山市的捷运系统与交通管理职能。旧金山有悠久的有轨电车历史，是一个空间紧凑、适宜步行与搭乘公交的城市。自1999年起，SFMTA整合了市区内的有轨电车、轻轨（包括地上与地下部分），出租车系统，自行车专用道系统，步行空间与景观系统，市区停车管理等多项功能。SFMTA大量运用低成本的交通改善措施，比如对快速公交系统专用道所处的地面进行粉刷标色，这一措施极大地提高了快速公交的运行效率。目前旧金山湾区是美国西海岸拥有最完善公共交通设施的城市，公交使用率在美国名列前茅，其交通状况与城市空间的质量都胜过同处一州的洛杉矶市。

2. 以香港为例，香港直到20世纪60年代才开始大规模的城市扩张。初期也一度轻视轨道交通建设，突出小汽车的拥有率。港英政府在60年代晚期开始制定完整的公交发展计划，在70年代的麦理浩时期开始大规模的基础设施与新城建设。目前港铁已经是一个拥有大量商业与住宅物业的经营实体，是世界上少有的盈利的公交系统。除了公交优先，香港对任何可能发生人车干扰的区域均进行空间、景观与信号标志系统的优化。尤其是香港在一定半径的轨道服务区内建立四通八达的步行系统，这一系统继承了明尼阿波利斯，卡尔加里等城市的类似系统，但是在轨道－物业整合的发展模式与产权结构的支持下，香港的行人友好性与系统的维护在亚洲乃至世界都居

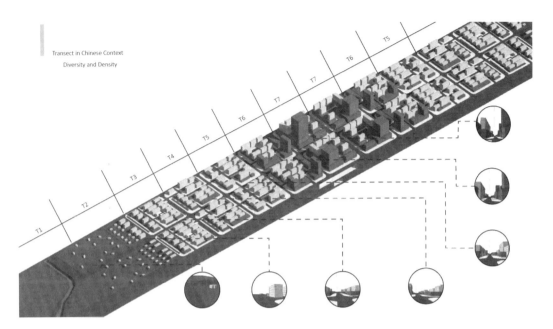

图5. 中国都会区社区的典型断面样例

设计应对雾霾 DESIGN AGAINST SMOG

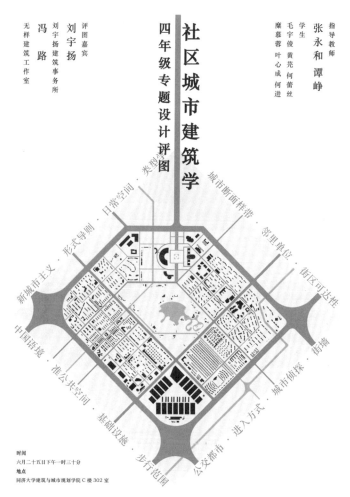

社区城市建筑学
四年级专题设计评图

指导教师 张永和 谭峥
学生 毛宇俊 黄芃 何蕾丝 糜慕蓉 叶心成 何进
评图嘉宾 刘宇扬
刘宇扬建筑事务所
冯路
无样建筑工作室

时间 六月二十五日下午一时三十分
地点 同济大学建筑与城市规划学院 C 楼 302 室

图 6. 笔者开设的基于新邻里单位模式的社区城市建筑学专题设计课海报（本课程已获得 2015 年度国家建筑学专业指导委员会教案奖）

图 7. 笔者及其研究小组总结的中国都会区的新邻里模式

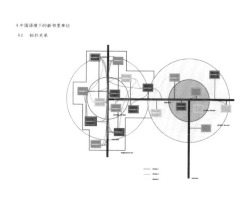

图 8. 网络社会的新邻里模型

前列。

通过比对，这些城市运用新城市主义进行交通改造的制度环境、地理条件与运用措施均不同，但是也拥有一些共同的特点：比如，都是在高速城市化的早期确立公交优先的发展模式；设立公交运营方与发展商利益共享的机制，把城市优化的成本分摊到在城市改造中得利最大的那些利益攸关方，即建立公交—物业的一体化制度设计；在任何可能发生交通拥堵的地区进行精细的道路设施设计；对行人专属区与商业区进行一体化设计等。许多城市都在立法层面对公交发展区制定了专门的发展规划。

中国式郊区

"机动性"是20世纪80年代以前的欧美规划学者追求的理想。他们一度认为，只要达到100%的私人汽车拥有率，就可以使得每个城市公民在择业、居住与日常生活中拥有完全的自主性。"机动性"概念经由现代主义建筑师休博赛默改造，成为树形的尽端式城市并推动着城市郊区化。郊区化即便在私人汽车占有率近乎100%的发达国家亦被证明是一种失败的规划策略。如果传统城市中心依然提供主要的工作机会，那么早晚潮汐式人口迁移不可避免。即使如加州湾区与洛杉矶地区那样的多中心式城市布局，每日早晚两次无序的人口流动也会使得整个高速公路系统瘫痪。与"机动性"相关的是超级街区概念，这种概念成熟于二战以后，即将数个街区连接为一个大街区，内部完全人车分行，并统一规划实现功能分区。这一概念在1949年前后就已经导入国内，而在1978年之后的城市建设中邻里单位渐渐演变为类似超级街区的"小区"，成为主导中国快速城市化的主要规划观念。

以往的观念是中国不存在"郊区化"这一美国独有的城市病症，但是这种理解混淆了郊区化的物质形态与郊区化的能量消耗方式。从物质形态来说，中国城市多以多层与中高层住宅小区为基底，较少低密度的独立住宅区，不存在郊区化的形象。但是从城市的物质与能量流动形式来看，广大城市边缘的新开发住宅区往往占据巨大的街区，内部严格限制非居住类功能，外部被快速路包绕，传统的街道生活被挤压。从城市体验来看，它们已经构成了蔓延式的标准化城市景观。从出行方式上来看，以上海为例，中外环一带的人口新导入区的职住分离现象严重，无论使用公共交通还是私人汽车均耗费大量时间，配套设施匮乏，社区生活缺乏，公共生活从传统的街道或集市转入远离社区的大型购物中心，事实上这些新开发小区或新城已经是一种郊区变体（Cevero & Day, 2008; Zacharias, 2005）。如果考虑整个城市化区域，那么中国三大城市带内部已经连为一体，农地与林地仅仅是点缀在巨大的城市化区域内的绿岛。

中国式的郊区化在中观层面上具有如下一些特征：

1. 由于我国的特殊的土地出让制度与开发商逐利本质的限制，城市的街区远远大于4~6公顷，边长200~250米的理想街区大小。

2. 城市干道缺乏中央隔离带或单行道二分路的形式，由于大街区的限制与封闭门禁的普遍运用，行人一出小区就会被迫横穿干道，或跨过没有任何无障碍设计或景观元素的天桥。

3. 即使某些地段被设计为步行街或行人专属区，也缺乏私人汽车—行人—公交系统的有组织的二维多高程设计，尤其是缺乏低成本的地面或高架公交走廊。

4. 公交与出租车停靠方式不合理，往往会与正常行驶的车辆发生冲突，或由于没有专用的公交车道，与其他车辆发生争道行为。

5. 都会中心区的高密度的商圈或金融区往往被历史上形成的城市干道穿越，或主街与干道不分，恶化了人车争道的状况。

6. 大众捷运与商圈物业缺乏整合，轨道交通服务半径内的设施不齐全。

新邻里模式

过往的研究多将城市的基础设施视为独立的机械系统,忽视社区形态与场所营造的热力学性能,往往将交通与物流等问题视为被动的适应城市扩张,并未主动减少激发无谓的交通因素,缩减城市升级的环境成本。对笔者来说,这一主动提升性能的关键在于重构邻里生活。本文通过百年的邻里规划的回顾探索从邻里行为方式的转变来提升城市性能的途径。必须注意的是,过去 30 年中国的土地供应方式与大地块的开发模式是违背城市自然生长规律的,大地块开发无法进行逐步城市更新,并将原本属于邻里层面(而非组团层面)的社区空间推给开发商,这决定了新城市主义所推崇的形式导则无法作用于邻里生活的塑造。形式导则的制定依然隶属于城市设计范畴而无法成为法定文件,或者必须依附于现有的控制性规划系统。在中国目前存量规划逐渐取代增量规划的情况下,要通过传统的形式导则的制定来引导社区发展愈加困难。形式导则或过于宽松以至于无法实现优化空间、降低能耗的目标,或过于苛刻而难以操作。

综上所述,中国规划语境下的新邻里模式研究至少面临两重挑战:第一是来自土地供应与开发方式;第二是来自规划框架。在挑战的同时,新邻里模式也有两大机遇。首先,改革开放之前及初期建造的大量工人新村已经开始老化,与旧式里弄与城中村不同的是,工人新村的初始格局符合邻里模式,可以通过局部改造来实现性能提升。其次,近年来,由于网络社交媒体的兴起,传统的商业街与购物中心开始转型,从传统的以购物主导的体验式、主题式社区转变。电子商务与社交媒体确实减少了人的出行需求,将需要空间迁移的社交活动转向虚拟空间,但是信息的流动往往与碎片化的物流形态同步发生,城市周边形成了巨大的物流服务带,物流成本上升,集成度降低,能耗虚高。这事实上是将个人的无序机动性转嫁为物流的无序机动性。这种能量与信息的无序活动是违背热力学第二定律的系统稳定原则的,所以网络社交媒体依然需要物质空间作为支撑。在这样的两重机遇条件下,急需建立一种新的邻里空间导则,即改变那种视邻里单位为封闭的热力学系统的陈旧观念,建立一种边界模糊,行为交互但又场所明确的社区形态。这一观念已经成为当代新城市主义发展的主流。

参考文献

[1] Jessop, Bob, Neil Brenner & Martin Jones. Theorizing socio-spatial relations // Environment and Planning D: Society and Space, 26 (3). 2008: 389-401.

[2] Johnson, Donald Leslie. Origin of the Neighborhood Unit // Planning Perspectives, 17. 2002: 227-245.

[3] Parolek, Daniel G., Karen Parolek. Form Based Codes: A Guide for Planners, Urban Designers, Municipalities, and Developers. Hoboken: Wiley, 2008.

[4] Robert Cervero, Jennifer Day, UC Berkeley Center for Future Urban Transport, University of California, Berkeley. Institute of Transportation Studies. Residential Relocation and Commuting Behavior in Shanghai, China: The Case for Transit Oriented Development. Institute of Transportation Studies, University of California, 2008.

图片来源:
图2,3,7,8:由本文作者提供。
图4:http://www.spur.org。

The New Neighborhood Model: The thermodynamics of new urbanism

Tan Zheng

Neighborhood and the Value of Localization

Modernization comes side by side with the development of mobility. Any imagination of 20th-century metropolises would have been related to highly developed transportation system, ranging from intercity railways to highways, from canalized transportation network to pedestrian overcrossing, from high-speed elevators to pedestrian conveyor belts. These technologies are all machines that promote the mobility of materials, labor and capital. Modernization is founded on the high energy consumption system on which such free mobility relies. Since the rising of the concept of Neighborhood Unit, it has been meant to resist high energy consuming mobility and to reconstruct the local thermodynamic value in the age of globalization. Such a value not only exerts its influence in the realm of energy, but is also relevant to spatial, social and cultural spheres. Through the exploration and renovation of several generations of urbanism scholars, the neighborhood model has been trying to create a compact and efficient community structure which optimizes the balance between work and life, and public service sphere, repressing the use of private cars and ultimately reducing meaningless and inefficient transportation volume and energy consumption on the level of inhabitant form.

Generally speaking, neighborhood means the orthodox term of the various historic forms of community in modern architecture and planning. In such a context, it doesn't provide more detailed rules for the form, and the organization or the scale of the community it refers to. In a narrow sense, the term Neighborhood Unit which has generated the modern concept of neighborhood represents a complete vertically coherent urban design system that involves principles, methods and practices. The term "neighborhood unit" was officially coined in the neighborhood design competition held by Chicago Urban Club in 1913. "Neighborhood unit" had become an unwritten principle popular with architects and planners during the 1920s. In 1929, the planner Clarence Perry provided a more detailed definition in the planning report Regional Plan of New York and Its Environs (Johnson, 2002). Originally it just created a pedestrian environment for school children. It has been generally developed into a self-perfecting living sphere spatial mode. Perry provided the six major principles for the neighborhood unit to comply with and drew a series of illustrations to describe the typical form of a neighborhood unit. In 1931, Perry's explanation of neighborhood unit was absorbed by Hoover government (1929-1933) and was gradually promoted with governmental housing policies.

During the process, it has become the standard model of neighborhood planning and real-estate property development. Its influence has lasted for forty years. In the neo-urbanism movement during the 1980s, the concept had been picked up again and become a significant source of theories in contemporary neighborhood design. In the last ten years, due to the deepening of globalization, the foundation of urbanology based on geographical proximity has been faced up with another challenge. The exploration of the form-based codes of neo-urbanism hits a bottle-neck (Parolek & Parolek, 2008). Under such a circumstance, we need to look back into the history of the response of the traditional physical space to search for the new pattern of neighborhood model in the thirty years of high-speed urbanization history.

Overflowing Mobility and Neo-urbanism

From the historical point of view, during the post-WWII period (mid-1920s), western developed countries had functionally districted cities according to the modernistic zoning concept, encouraged the separation of working and living areas, indulged the overflowing development of private vehicles. The development mode represented by highways, elevated roads, motorways and superblocks that had dominated the urban infrastructure updating in Chinese cities in the 1980s and 1990s. It partly advanced the urban transportation efficiency, but in the meantime sacrificed the traditional small blocks that formed in history and deteriorated the traffic environment of people and non-motorized vehicles. Since the 1980s, urbanologists in America have introspected the issue of environmental deterioration due to the spread of highways. A lot of new planning theories aimed at solving suburbanization problems emerged during the period. The new urbanism theory based on the neighborhood model was one of them. The narrow theory of new urbanism mainly aims to compensate for the block scale and neighborhood relationship lost during the process of urban expansion, and advocating traditional (European) block form and localized lifestyle. General new urbanism has created the public system oriented integrated form guided principle of region-town-neighborhood-block-architecture and optimized the existed public space, road structure and landscape system to change the traffic modes of civilians. On macro level, new urbanism's interpretation of regional city conforms to the theories of contemporary economic geographers such as the theory of Neil Brenner and that of Bob Jessop. On the metro level, new urbanism conducts the derivation of ideal urban density and type with the method of "urban fracture surface sample", etc. On the micro level, new urbanism advocates architectural control guide principles that conform to cultural inertia and form cognition. Generally speaking, new urbanism is a synthesized knowledge system. Other urban theories and environment cognitive theories of the same period still ignored the post-positivism science and philosophy of Thomas Kuhn, presumed researchers that were completely isolated from objects and ignored the fact that urban theories were summarized from practice.

Although there still exists some criticism however, undeniably, new urbanism has provided the most productive strategies and modes in contemporary urban practice. New urbanism has exerted great influence on the urban design practice in contemporary America and directly guided the organization of many rules of neighborhood forms. New urbanism deals with regions that have completed urbanization (suburbanization) through minimum environmental intervention and relatively high investment efficiency to increase the density of cities, reduce energy consumption and vehicle traffic and improve public space. Presently the process of urbanization in China has shifted from high-speed development stage to medium-speed development stage and transformed from the "increment planning" level to the "stock planning" level. Many metropolises have hit the land supply bottleneck. The existing extensive-style development has become unsustainable. The original intention of new urbanism was to reconstruct the built urban (suburban) regions overwhelmed by private vehicles. It has become a great source of reference for the current planning situation of our country. It is worth noticing that many new urban construction projects with the name of "new town" wave the flag of "new urbanism" and narrowly defines new urbanism as some architectural styles and features, but differs a lot from the principles of new urbanism in detailed planning methods.

Successful Cases of New Urbanism and Public Transportation Cities

New urbanism and the concept of public traffic oriented development zone (TOD) was first worked out to target the maladies of urban planning in America. Later it gradually spread the influence to Singapore, Hong Kong and other high-density Asian cities. Presently, it is generally acknowledged by the planning academic circle that cities in North America and Asia that comply with the standard of public traffic city include Washington D. C., Seattle, San Francisco, Calgary in Canada, Singapore and Hong Kong. Here are two examples.

Take San Francisco for an example, San Francisco Municipal Traffic Administration controls the whole Muni system in San Francisco and traffic management functions. San Francisco has a long history of electric tram cars. It is a spatially compact city that is suitable to walk around or travel in public vehicles. Since 1999, SFMTA has integrated the tram car system, the light railway system (including the above-ground and underground parts), the taxi system, the bicycle lanes system, the pedestrian space and landscape system as well as the downtown parking administration system etc. SFMTA applies a great number of low-cost traffic improvement methods such as signifying transit-only lanes with whitewash. The method has greatly increased the efficiency of high-speed public traffic. Presently, San Francisco Bay Area is one of the cities that has the most excellent public transportation facilities along the western coast of America. San Francisco is at the top of the list in public traffic utilization ratio. The traffic condition and the quality of urban space in San Francisco are better than Los Angeles in the same state.

Hong Kong is another great example. The city had not started the process of high-speed expansion until the 1960s. During the early days, the city neglected the construction of railway traffic system and focused on the owning rate of private cars. The British Hong Kong government issued the public traffic development plan in the late 1960s. In the 1970s, under the administration of MacLehose government, Hong Kong started large-scale infrastructure construction and new town development. Today MTR is a business entity with a lot of business as well as residence properties. It is one of the few profitable public systems in the world. Hong Kong not only bestows public traffic with priority, but also optimizes any area that may initiate human-vehicle intervention with the improvement of space, landscape and signal system. Noticeably, Hong Kong has established a pedestrian system accessible from all directions in the catchment within a certain radius. The system has inherited the similar structures in Minneapolis, Calgary, etc. However, with the railway-property integrated development mode and the support of property right structure, Hong Kong is still at the forefront in its pedestrian-friendliness and the maintenance condition.

By contrast, the cities differ from each other in their application of New Urbanism in traffic pattern, political environments, geographic conditions and practice methods. However, there does exist some common features. For example, the cities have established the public traffic oriented development model during the early days of high-speed urbanization. They have also established the system of sharing interest among public traffic operators and developers to distribute the cost of urban optimization among the interests that profit most greatly in urban renovation which means constructing a public traffic-property integrated system. These cities have conducted meticulous traffic facility design in areas where traffic jams tend to occur. They have also employed integrated design for pedestrian areas and business areas. Many of these cities have come up with particular development plans for public traffic development areas on the level of legislation.

Chinese-style Suburb

"Mobility" used to be an ideal pursued European and American planners and scholars. In their opinion, as long as the rate of car-ownership reached 100%, every citizen would have full autonomy in choosing profession, inhabitation and daily life. The concept of "mobility" has been modified by the modernist architect Ludwig Hilberseimer to be a tree-structured out-de-sac-style city model and promotes the suburbanization of the city. Even in developed countries with close to 100% car-ownership rate, suburbanization has been proved to be a failed planning strategy. If traditional city centers still provide the majority of work opportunities, tide-like transfer of population is unavoidable. Even in multi-center cities such as California Bay Area and Los Angeles, chaotic population transfer every morning and every evening would still paralyze the whole highway system. The concept "super neighborhood" is related to "mobility". The concept had been developed during the post WWII period. It means connecting many blocks into a large block. Within the block, people and vehicles are completely separated. The whole block is planned and divided into functional sections. The concept has been introduced into China around 1949. After 1978, the neighborhood unit in urban construction has developed into "microdistricts" similar to super-blocks and become the major planning concept in the fast urbanization in China.

It was considered that the "suburbanization" was an urban symptom exclusive to America. However, the view obscured the physical form of suburbanization and the energy consumption form of suburbanization. From the perspective of physical form, the basic architectural units in China are mainly multi-storey buildings and middle or high residential neighborhoods. There are few independent low-density neighborhoods. There's no sign of suburbanization. However, from the viewpoint of urban material and energy transfer, newly developed neighborhoods on the edge of the city often occupy a large part of the block. The interior of the space strictly limits non-living functions while the exterior is enclosed by high-speed roads. The traditional street life is compressed. From the perspective of urban experience, they have constituted the standard expansive urban landscape. From the perspective of transportation, taking Shanghai for an example, the newly developed neighborhoods around the middle ring and the outer ring have led to the separation of working and living functions. Public life has transferred from traditional streets or markets to large shopping malls far from the neighborhoods. Actually these newly developed neighborhoods or new towns have become a variant of the neighborhood (Cevero & Day, 2008; Zacharias, 2005). If we take the whole range of urbanization into consideration, the largest three city zones in China have been connected with each other internally. The farm land and the timber land are just green islands interspersing the huge region of urbanization.

On the meso-level, Chinese style suburbanization have several features:

Due to the special land grant policies and the limitation of the profit-driven nature of developers, a typical urban neighborhood is much bigger than the size of an ideal neighborhood which measures 4~6 hectares with the size length of 200~250 meter.

As the main streets in the city lack central isolation belt or the one-direction-two-way form, and due to the limitation of large blocks and the universal employment of strict entrance guard, pedestrians are forced to cross the main street or an overline bridge without any barrier free design or landscape elements.

Even some areas have been designed as pedestrian malls or exclusive zones for pedestrians and lack the organized 3D multi-level elevation design consisting of private car-pedestrian-public transportation system, especially in low-cost ground or elevated public transportation corridors.

The unreasonable parking way of buses and taxis often leads to conflicts with moving vehicles or cutting in because there are no exclusive public lanes.

High-density business circles or financial areas in city centers are often cut across by historically formed arterial roads. Or mixed main streets and arterial roads deteriorate the situation of cutting in.

Public metros are not integrated into business circles. The infrastructure within the radius of railway transportation service is incomplete.

The New Model of Neighborhood

Past researchers tended to consider the urban infrastructure as an independent mechanical system and ignore the form of the community or the thermodynamic features created by the site. They also tended to consider issues of transportation and logistics as passive elements that conformed to the expansion of the city. Therefore, they rarely actively reduced elements that would stimulate unnecessary traffic or cut down the environmental cost of urban upgrading. As far as the author is concerned, the key point of performance promotion lies in the reconstruction of neighborhood life. This essay tries to explore the way to enhance the performance of the city through the transition of neighborhood activities with a retrospective of the last hundred years of neighborhood planning. It is worth noticing that in the last thirty years the methods of land supply and the development pattern of large land parcel in China have violated the natural growth law of the city. The development of large land parcels cannot realize urban renewal gradually. Community space used to belong to the neighborhood level (instead of non-group level) has been transferred to developers. Therefore formal guidance promoted by new urbanism can't exert its influence on the shaping of neighborhood life. The constitution of formal guidance is still subordinate to the realm of urban design which means it can't be accepted as legislative documents. Or it must rely on existing administrative planning system. In today's China, as stock planning is gradually replacing increment planning, it is more difficult to guide neighborhood development with the traditional constitution of formal guidance. The formal guidance is either too loose to realize the objective of optimizing space and reducing energy consumption or too rigid to practice.

In conclusion, the new neighborhood model in the context of Chinese urban planning at least faces two challenges. The first challenge comes from the way of land supply and development. The second comes from the planning frame. Besides challenges, the new neighborhood model also has two great opportunities. Firstly, a great number of workers' neighborhoods around the issuing of the reform and opening policy are getting old. In contrary to old lanes and alleys or urban villages, the original layout of workers' neighborhood accords with the neighborhood mode. Hence it is possible to realize functional improvement through partly renovation. Secondly, due to the rise of social media in the last few years, traditional business streets and shopping malls are starting to transform from traditional shopping-oriented venues to experience and theme oriented communities. Electronic business and social media have actually reduced the traffic demand of people and turn social activities that require spatial transportation into virtual space. However, the transfer of information often occurs simultaneously with fragmented logistic form. A huge belt of logistic service has formed on the edge of the city. The cost of logistics is increased. The integration level is reduced and the energy consumption is unnecessarily high. Actually, the unordered activities of energy and information violate the system stability principle of second law of thermodynamics. Therefore social media still needs physical space as its support. Under such a condition, it requires to constitute a new neighborhood spatial guidance to alter the old concept that considers neighborhood unit as enclosed thermodynamic system and establish a boundary blurring, activity interactive and site specific form of community. The concept has become the mainstream direction of the development of contemporary new urbanism.

建筑废料
纪念碑

刘宇扬

城市雾霾和其他环境问题是否由于建筑而生？又是否能从已知的建筑体系中找到解答？这是我和我的教学搭档高亦陶老师在今夏的同济"设计应对雾霾"工作坊中所尝试思考的议题，也是我事务所ALYA团队与英国Chora团队在近期刚完成的一个合作项目中——江桥滨江爱特公园所试图提出的设计回应。

当我们在看城市雾霾问题时，我们看到的不是单一的基地或建筑功能所引起的问题。实际上，建筑师所需要面对的是如何通过重新在概念上建构一种平行而又互补的环境交换系统，来回应雾霾和其他环境问题。这种交换可以被理解为不仅是环境问题的症状和原因，同时也是环境问题的生产和消费。环境交换的双重性形成了非常重要的一点，就是对抗环境问题的设计策略不再局限于特定的某栋建筑物、场地、或文脉类型，而是通过一个更加广泛的空间、社会、及生态系统所建立的建筑学，在此同时也重新界定了"建筑学"对未来实践的意义。

在课程设计中，我们选择了陆家嘴与江桥镇做为两个对照组。这两个地方也可以同时理解为由苏州河连接的两点一线。两个地方互为头尾，却说不清哪个是头哪个是尾。在苏州河入黄埔江的这端，作为中央商务区的陆家嘴，可以说是饱受来自上海周边工业聚集区域的雾霾之害；而高楼耸立和有着大量办公人群聚集的陆家嘴，却同时消耗了大量的能源并制造了大量的垃圾，对周边如位于苏州河上游（称之为吴淞江）边上的江桥镇这样的工业地区，造成巨大的环境压力。

因此，在环境问题上，我们可以说两个地方同时互为因果关系。而我们希望探讨的设计提案是一种对雾霾与环境问题中双重特性的协作回应、研究和批判。第一阶段的研究是针对一系列特定主题的原型开发，陆家嘴组的原型研究主题是关于典型高层建筑的三个基本元素：表皮、核心筒、大堂，江桥组的原型研究则是围绕着废弃资源回收主题中的三个基本概念：捕捉、处理、转移；第二阶段的设计需要同时针对两个地块并综合以上的两组原型，形成对比方案。学员们在原型研究过程中，都能得到一些有趣和出乎意料的发现，特别是资源回收的种类和处理模式。

遗憾的是，团队在短短两周的时间里，未能充分发挥前半段原型研究的基础与启发，同时个别学员的思路相对封闭保守，最终得到的提案停留在传统英雄式的形式主义和空间手法的层面上，而非更具系统性的动态策略，可说是远远未能达到我们原本预期的实验性设计成果。从实践视野来看，这个结果其实对目前主流建筑价值观提出了一个严肃的反思和批判的机会。多数的学生和实践者，不论是在中国还是西方，从根本上都没有意识到这个行业的专业训练与我们的环境与自然生态是多么的割裂，而所设计出来的，不论是产品还是所谓作品，不仅对环境无任何助益，而实际上制造了一大堆"建筑废料纪念碑"，并持续不断地对环境进行破坏。然而，这个工作坊的安排，至少迈出了改变思想的第一步。

在我与柏林工大的拉乌尔·邦休顿教授合作设计的上海江桥滨江爱特公园项目中，我们试图重新定义了"建筑废料纪念碑"。

在过去数十年的快速城市化进程中，特别是老城和农村动拆迁过程中，建筑废弃物的处理是一个不为建筑行业所关注的问题。首先，国内相

关规范对建筑废弃物的定义是建设单位、施工单位新建、改建、扩建和拆除各类建筑物、构筑物、管网等以及居民装饰装修房屋过程中所产生的弃土、弃料及其他废弃物。中国目前对建筑垃圾的处理大体可以分为两类：一是将建筑垃圾进行轻度分拣，回收废金属、废混凝土等，采用这类处理方式的建筑垃圾仅占2%；二是未经任何处理的建筑垃圾被运到郊外或者农村，买或者租块地，采用露天堆放或填埋的方式进行处理，采用这类处理方式的建筑垃圾约98%。

我们的基地位于居民区与工厂交界的灰色地带；土地上堆放着大量建筑垃圾及土方，杂草丛生；周边道路来往着工程车辆，灰尘滚滚，居民们都对这块"废地"避之不及。我们的设计过程中本着上方原地消化及再利用的可持续理念，创新地利用废弃建筑垃圾填充石笼，形成标志性的地景构筑物。既留住了基地的过去，又活化了"废弃空间"成为周边小区的公共社交空间。

最南侧的石笼墙体长202米，高5米，既是贯穿整个公园的标志性景观，也是阻隔周边水泥厂污染的屏障。公园内部由350米长的健身步道串联起入口广场、健身广场、儿童活动区和各个石笼构筑物。花草和乔木搭配形成四季变化的景观，成为具有亲和力的社区公园。

实际上，在介入这个公园的具体设计之前，刘宁扬建筑事务所已在2014年与深圳的都市实践、上海的集合设计和思锐设计团队合作，由当地政府委托对吴淞江北岸沿江范围进行过一轮的概念性城市设计。这个地方属于一个所谓城乡结合部。由于大量的城市开发，造成了很多的建筑垃圾，在很长一段时间内堆放在这个场地上，没有一个具体的政策甚至预算去处理这样的一个问题。它的南边还有一个相当大的水泥厂，形成一个严重的污染源。由于前期的基础工作，让我们

团队对周遭的环境问题已经有了比较深刻的了解，而我们的城市设计提案的着重点也恰恰对环境治理问题方面提出一些土壤修复和水质净化的建议。

而项目的起因也就在于居民对这个现象已经开始不满，对地方政府形成一定的压力。其实一开始大家想的仅仅是如何把这些废土废料移到另外一个地方。设计就从这儿出发，如何解决好场地上的废土废料，第二如何能够利用这个地块做一个天然屏障把水泥厂的粉尘做有效的阻挡。当我们在现场的时候，我们发现到周边的小区已经形成一定的社区关系：居民和老百姓需要经过这么一个垃圾堆，要呼吸着这个水泥厂所散发出来的粉尘。我们认为设计上应该要还原给居民一个干净的休闲空间，一个健康的社区空间。

我们的策略是尽量能就地把废土重新去整理重新去利用。借用了在水利工程中经常用的生态石笼，把现场所有能够搜集起来的石头和砖全部往里面放，形成一个体量比较大的景观构筑物，我们称之为废料纪念碑。在构筑物之间我们种上树，并设计了一些社区的活动设施，有的地方留白，让社区的人能够享受这样社区环境。

在两百米长的从公园的一端到另外一端的生态石笼墙中，我们预留了三个通道，做为与二期的河滨公园的链接空间。除了墙以外，我们还用同样的方法去堆积成平台，其中的一个是居民可以走上去的。这个平台成为一个制高点，让人回顾这些在很短的建设历史中所创作出来的废料纪念碑。

现场有超出原本预期的废土量，比较大的问题是关于土方平衡，使得我们必须不断地去调整场地的设计标高。而由于若干年的堆砌，原有场地也已形成了一种几乎是自然地景的现象，甚至里面有个小山包，上面还有一棵树，居民都喜欢爬上去。所以在我们的设计上，最后一个动作是把这个山包保留下来了，原来的树在施工过程中不幸被推倒，我们便种回更多的树，同时，在山包脚下的小广场当中设计了一个小亭子，用竹钢跟阳光板形成一种更轻盈的构筑物。

在设计上还有其他许多相当丰富的小细节，从水洗石收边到步径上的材料与灯光。在景观的选择上尽量选择较大的乔木，形成一些阴凉的小环境；在公园的座椅设施上，尽量自行设计然后由工人来现场制作，某种程度其实也是对于传统工匠手艺的致敬，在当下社会中，许多设计的配套设施都是直接购买现成品，我们更愿意把它设计出来并借用匠人的手做出来，希望居民在使用的时候，能够感受到设计对人的关怀。

Memorial of Building Remains

Liu Yuyang

Urban smog and other environmental issues are caused by architectures? Can we find a solution from the known architectural system? My teaching partner Oscar Ko and I tried to contemplate on the issue with the Tong Ji summer school workshop "Design against Smog". It is also the design response proposed by a cooperative project, Ai Te Riverside Park in Jiang Qiao by the ALYA team of my firm and the British team Chora.

When we approach the issue of urban smog, what we see is not a problem caused by a site or program. In fact, architects need to figure out how to constitute a concept model in which parallel and reciporcal environmental exchange system is designated to response to smog and other environmental problems. This exchange can be understood not only as the symptom and the cause of environmental problems, but also as the production and consumption of environmental problems. The duality of environmental exchange has addressed a crucial point, that design strategies are no longer limited to specific buildings, locations or context types. The architecture established by a broader spatial, social and ecological system redefines the meaning of "architecture" to future practice.

We chose Lujiazui and Jiang Qiao Town as two control groups in the course design of our team. The two locations can be understood as two dots connected by Su Zhou River, facing each other across the river. But it is hard to tell which one is the head or which one is the end. As a central business area, Lujiazui is located at the end where Su Zhou River joins Huang Pu River. It suffers a lot from the smog created by the industry zone on the periphery of Shanghai. At the same time, the countless high-rises and business groups consume huge amount of energy resources and create a lot of waste. It has exerted great environmental pressure on the industrial areas along the upper reaches of Su Zhou River (called Wu Song River).

Hence environmental problems considered, these two locations interact as both cause and effect. The design proposal we put forward is a comprehensive response, survey and critique towards the duality that vesicles in smog crises. The study of the first phase is aimed at the paradigm development of a series of certain themes. The subject of the paradigm study of the Lujiazui case is related to three fundamental elements of high-rise buildings: skin, core tube, foyer. The paradigm study of Jiang Qiao case is centered on the three basic concepts of waste resource recycling: capture, process and transfer. The design of the second stage requires targeting the two land parcels at the same time and integrate the two groups to form a comparative plan. Students can make some interesting and surprising discoveries during the paradigm research, especially the types and processing modes of resource recycling.

Regrettably, the team didn't manage to fully exert the foundation or inspiration of the first phase of the paradigm research. In the meantime, due to the relatively conservative way of thinking of a few students, the final proposal settled on the level of formalism and spatial measures of traditional heroism instead of more systematic and dynamic strategies. It is far from realizing the original design of the experiment or the achievement we predicted.

设计应对雾霾 DESIGN AGAINST SMOG

From the perspective of practice, the result actually put forward an opportunity of serious retrospect and criticism of the present mainstream values of architecture. Most students and practitioners, Chinese or westerners have not realized how the professional training of the business is isolated from our environment and natural ecology. The outcome of the design-products or works makes no contribution to the environment and even creates a lot of "architectural garbage memorials" that continuously deteriorate the environment. The arrangement of the workshop at least takes the first step to change the way of thinking.

In the project of Ai Te Riverside Park in Jiang Qiao, Shanghai, cooperated with Professor Raoul Bunschoten from Technische Universit't Berlin, we tried to redefine the "architectural garbage memorial."

In the fast urbanization process during the past ten years, the processing of architectural garbage is an issue neglected by the architectural world, especially those come from the relocation of old towns and villages. First of all, related rules in China defines architectural garbage as discarded soil, discarded material and other waste created by construction companies, building companies in the process of building, reconstructing, expanding and demolishing various kinds of buildings, structures, pipe networks and those produced by residents in the process of home decoration. Presently, the processing measures of architectural garbage in China are divided into two categories. The first is to roughly categorize architectural garbage into, which only constitutes 2% of the total. The second is to transport unprocessed architectural garbage to the suburb or countryside, buy or rent a land and open-pile or bury the garbage. The method accounts for 98%.

Our site is located in the grey zone at the border between the residential area and the factory, with a lot of architectural waste and earthworks piling up, overgrown with wild grass. Dense dust swirled and billowed as trucks drive by. Local residents all try their best to avoid the "wasteland". In the process of our design, we bore the sustainability idea in mind that earthworks could be consumed on site and reutilized, innovatively utilized discarded architectural garbage to fill gabions and create symbolic landscape structures. The design not only maintained the past of the site, but also activated the "waste space" and transited it into a public social space of the surrounding neighborhoods.

The southernmost gabion wall is 202 meters in length and 5 meters in height. It is a landmark running through the whole park as well as a barrier shielding against the pollution of the neighboring concrete factory. In the park, the entrance square, the fitness square and the children playground and gabion structures are connected by a 350 meters long step way. Flowers and trees match with each other to form the view changing with seasons. It is a cozy community park.

In fact, before participating in the detailed design of the park, the architectural firm of Liu Yu Yang has cooperated with design teams Urbanus based in Shenzhen, One Design inc. and Serie architects in Shanghai on a round of conceptual urban design project along the north bank of Wu Song Jiang commissioned by the local government. The location belongs to a so-called rural-urban fringe zone. Due to large scale urban development, there is a great amount of architectural garbage piled on the site for a long time. There is no specific policy or even budget to deal with such a problem. Also, a large concrete factory on the south forms a serious resource of pollution. Because of fundamental work of the first stage, our team had got a profound comprehension of the surrounding environmental problems. The urban design proposal that we put forward thus focused on advices of soil remediation and water purification on environmental administration issues.

The reason why we started the project in the first place was because local residents had started to be discontented with the situation and exerted a certain extent of pressure on the local government. In fact, at the beginning, the removal of discarded soil and material become our first concern and departure of design, along with how to ues the plot as a natural barrier. We just hoped to figure out a way to move the discarded soil and material to another location. It was also the starting point of our design to deal with the discarded soil and material on the site and to use the land parcel as a natural barrier to effectively shield against the powdery dust emitted by the concrete factory. While conducting field study, we found that surrounding neighborhoods had already formed a certain communal consensus: residents and civilians need to pass the garbage pile and breath in the dust emitted by the concrete factory. We thought the design needed to restore a clean leisure space, a healthy community space to the local residents.

Our strategy was to try our best to reorganize and reutilize discarded soil on site. The idea of ecological gabion was inspired by the ones used in irrigation words, and gabions that contain the stones and bricks collected from site form a view structure relatively big in volume. We call it a garbage memorial. Then we planted trees on the structure and designed some activity equipment for the community. Some part was intentionally left blank to enable the residents to enjoy such a community environment.

We left three passages in the 200-meter ecological gabion wall ranging from the end of the park to the other end as the linking space to the phase two riverside park. Besides the wall, we created several platforms with the same method. One of them was accessible to residents. This platform has become a high point enabling people to retrospect the garbage memorial created in the short history of construction.

There was more discarded soil than expected and balance of earthworks was a big problem. We were forced to constantly adjust the standard design height of the location. Due to several years of piling, the original site had formed an almost natural landscape. There was even a small hill with a tree. Many inhabitants like climbing it. Therefore we kept the small hill in our design at last. Unfortunately, the original tree was pushed down during the construction. Hence more trees were planted. In the meantime, we designed a small pavilion for the little square at the foot of the hill, a lighter structure made of bamboo steel and sunshine plates.

There were also a lot of rich small details in the design, from washed stone edge to the material and the light of the pedestrian trail. We tried to use bigger trees to create shades. We also tried to apply original and site-specific design to benches and other equipment in the park. To some extent it was also a tribute to traditional craftsmanship. In contemporary society, a lot of supporting facilities of the design are readymade. On the contrary, we preferred original design and realized it with the hands of craftsmen. We hoped that the residents could feel the care of the design for humanity.

步骤：对社区可持续性问题的系统性思考

苏运升，
乔治·吉奥吉夫

在为期两周的暑期课程"设计对抗雾霾"中，两位应用研究顾问和由六名学生组成的国际团队作为一个跨学科小组进行合作，通过理解高度工业化的城市社会的复杂性和诸多挑战而创造了一个系统性的观点，并且为其寻找改良的解决方案。

通过为先进产业4.0服务提供一个系统性平台，将重要利益相关者、技术、社会经济原则整合在内，这一创新团队建立了新一代市场途径的核心。我们可以将整个城市有机体理解为一个由几个因素构成的模块体系：

- 时间
- 空间
- 技术
- 人
- 金融
- 规范
- 政策
- 热力学
- 环境
- 空间秩序

该团队提出了一种服务于城市结构及其公民需求的整体工具，整合了各个科学领域的知识：

- 建筑
- 城市规划
- 工程学
- 地理
- 社会学
- 人类学
- 经济学
- 政治学
- 环境工程
- 信息科技

这种高强度工作的成果可以描述为一种软件工具，首先它应是一种教育性质的软件，有可行性、结构及界面对用户友好。

在这个服务工具的核心有一个平台，连接了所有社区开发利益相关者，通过引入金融和规范控制机制，提供稳定的服务质量，一整套服务体系和易于预测的工作流程及工作时间。

在当前社会，城市规划在传统工业化的进程中起到一种线性作用。目前它的效率非常低，而且无法融入持续崛起的社会创新洪流。

当前的方法尝试推出一种互动"游戏"平台，它使用多种交互原则，以便探索城市整体发展的可能性。社会网络创新平台，应该将城市建造和更新引向明晰的协同和规范的公私合作。

这个游戏平台中明晰的知识产权、物权和空间所有权构成了它的基础。财产所有权的使用非常明确，可以呈现出多重视角而让这个平台更强大。通过多种社会角色的协调，它旨在提高空间发展的效率和决策效率。其最终目标是：创造大量积极的社会效应和附加效益。

平台工具的核心功能如下：

1. 需求。可以描述与新基础设施相关，或对

设计应对雾霾 DESIGN AGAINST SMOG

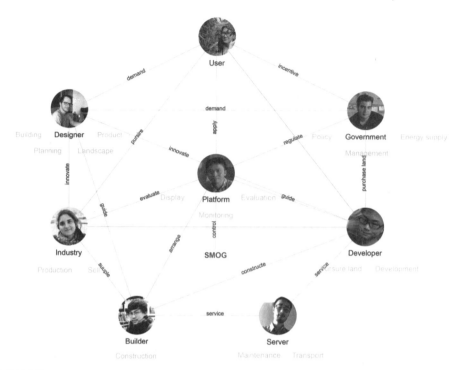

图表 1. STEP 的工作机制

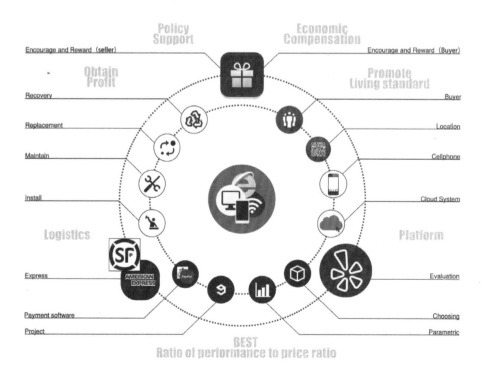

图表 2. PEST 元素

于现有基础设施翻新的各种需求。

2. 平台。团队融合了多功能的专门知识，社会人际网和规范工具，确保各个方向的安全工作流。

3. 服务。将各种服务整合成一个类别——建造、翻新、维护、拆除、回收、二手市场和价值服务等。

4. 相关利益者。相关利益者——开发商、业主、政府、产业、市场、质量控制部门等。

从 PEST 到 STEP——能效 + 对抗雾霾的可再生能源产品和服务平台

雾霾是一个地区性的问题，是由对化石能源不加限制而与日俱增的使用造成的，这种能源已经被人类社会作为基本的主要能源集中大量使用了长达 150 年。设计对抗雾霾是一种将人类能源消费结构从矿物资源重新转向可再生资源的手段。

STEP 提案并不是只针对陆家嘴的设计选择，而是利用陆家嘴的声誉作为一种推广平台，把一种集成式的解决方案传播开来，作为一种产品和服务组成的纲要，以便加速将能源消费结构向平衡而可持续的社区有机体转变。

在工业革命之后，基于城市能源结构从可再生能源向化石能源的转变过程，现代城市规划理论诞生了，主要是缘于与日俱增的水平和垂直定居密度。于是现代城市规划理论"指引"人类依赖化石资源的使用，而非鼓励可再生能源的使用。但在现今的城市和建筑规划进程中，大多数规划者应该关注如何提高能效。

项目的缩写包含了四个 STEP 机制——社会意识、技术创新、经济逻辑、支持和政策，这四个机制一同构成了一种灵活的多主体机制，协调了责任、权利和利益。

STEP 机制，与传统的 PEST 进程不同，将政策、协商和经济支持置于最前。STEP 进程试图鼓励社会意识和技术创新，以此作为起点。最终目标：建立起互联网平台，匹配空间所有权和知识所有权，最终创造附加效益。

从 STEP 设计进程来说，该平台提出了一种机制协调能源生产和节约贷款、金钱和能源，建立在空间、城市和建筑规划政策基础上。与由于缺乏知识和经验而仅仅依赖资金基础，不要求系统性手段的传统规划过程相比，这个机制非常优越。从 STEP 城市发展进程来说，封闭和排外的决策过程被开放而充满活力的信息系统代替。基于这一点，STEP 设计进程系统能够刺激能效优化产品和服务，以及促进可再生能源产品和服务的推广。

第七小组的创意提出了一个全新的基于互联网的名为 STEP 的建筑和城市规划框架，它提供转变能源消费、生产结构的解决方案的条目，同时结合了管理机制、设计过程、建筑过程和质量控制。除此之外，为了吸引更多年轻人的目光，第七小组将这个平台的起点设计为一个电脑游戏——亦是一个教育工具。

STEP: A Systematic Consideration of Settlement Sustainability Issues

Su Yunsheng, Georgi Georgiev

During the two weeks of the Summer School 'Design against smog" two applied research advisers and an international team of six students did work in an interdisciplinary team, creating a systematic idea pool by understanding the complexity and challenges of highly industrialized urban societies and searching for improvement solutions for those.

By creating a systematic platform for advanced industry 4.0 services and introducing significant stakeholders, technologies and socio-economic principles, the innovation team established the core of a new generation market approach. To understand the overall urban organism as a modular system comprises of following factors:
- Time
- Space
- Technology
- People
- Finances
- Regulation
- Policy
- Thermodynamics
- Environment
- Spatial order

the team proposed a holistic tool for serving the urban fabric and the needs of its citizen, combining knowledge in various areas of science:
- Architecture
- Urban planning
- Engineering
- Geography
- Sociology
- Anthropology
- Economy
- Policy
- Environmental engineering
- IT

The result of the intensive work could be described as a software tool, which firstly should be developed as an educational software, in order to prove its feasibility and user friendly structure and interface.

In the core of this service tool, there is the platform, connecting all settlement development stakeholders, by introducing financial and regulation control mechanisms, providing a continuous service quality, a full range of services and easy-to-predict workflow and work time.

In the present society, the urban planning is a linear function of the traditional industrialization process. It is currently very inefficient, and cannot be incorporated into the way of the continuously rising social innovation.

The current approach attempts to raise an interactive "game" platform, using multiple interactive principles, in order to explore the possibility of comprehensive improvement of urban development. The construction and renewal of the city towards the social networking innovation platform should lead to a clearly coordinated and regulated public-private partnership. Clear intellectual property and space ownership in this game platform describe its fundaments. The use of property rights is clear, and can make the platform powerful from multiple perspectives. Through the coordination of multiple social roles, it aims to improve the efficiency of spatial development and the efficiency of decision making. The final goal is to create a large number of positive social effects and additional values.

The core function of the proposed platform tool could be described as follows:

1. Needs – as the complete variety of needs regarding new infrastructure or the retrofit of the existing one can be described.

2. Platform – in the team incorporated a multifunctional know-how, social networking and regulatory tool, which assures a secure workflow in every direction.

3. Sevrvices – unifies as a category, the entire variety of services – construction, retrofit, maintenance, dismantling, recycling, second – hand market, add value services etc.

4. Stakeholders – developers, owners, government, industries, market, quality controllers etc.

From PEST to STEP: the Energy Efficiency + Renewable Energy Products and Services Platform against SMOG

Smog is a regional problem, and is caused by the unlimited rising use of fossil energy, which has already been used in extreme amounts as the basic main energy by human society for the last 150 years. Design against Smog for an approach to transform the anthropogenic energy consumption structure from fossil sources back into renewable resources.

The STEP proposal is not a design option only for Lujiazhun, but using Lujiazu's fame as an advertising platform, to spread out a micro integrated solution, as a compendium of products and service components, in order to fasten the conversion process of the energy consumption structure towards a balanced on sustainable settlement organism.

After the Industrial Revolution, urban planning theory of the modernism was born, based on the transformation process of city energy structure from renewable energy to fossil energy, mainly because of the rising horizontal and vertical settlement density. So the modernist urban planning theory "Has brought" human beings to depend on the usage of fossil rather than encourage the usage of renewable energy. During the process of urban and building planning nowadays, most planners should concentrate on increasing the energy efficiency.

The four STEP mechanisms are delicated in the abbreviation — Social Awareness, Technology innovation, Economic logic/support and policy, and form together a flexible mechanism of multi-subjects coordinating responsibility, rights and interests.

The STEP mechanism, different from the traditional PEST process, put the policy/negotiation and the economic support in the front. The STEP process tries to encourage social awareness and technology innovation as the starting point. The final goal: building up internet platform, which matches space ownership and idea ownership, to finally create an added value.

In terms of the STEP design process, the platform suggests a mechanism to coordinate energy production and saving loan, money and energy, based on the spatial, urban and building planning policy. It is superior, compared to the traditional planning process, which only relies on the monetary basis and not requiring a systematic approach, mainly because of the leak of knowledge and experience. In terms of the STEP city development process, the closed and exclusive decision process has been replaced by an opened and dynamic information system. Based on this, the STEP design process system is able to stimulate the promotion of energy efficiency optimization products and services, renewable energy products and services.

The idea of Group7 proposed a new STEP internet-based building and urban planning framework, which should provide solution catalog of transforming the energy consumption/production structure, combining management mechanisms, design process, construction process and quality control simultaneously. In addition, in order to attract more attention by the young audience, Group7 designed the platform's start as a computer game – an educational tool.

城市地区空气污染现状和研究方法

李卓,陶文铨

空气污染是由于颗粒物、化学和生物分子改变了大气自然特性而产生的后果。空气污染物质可以是固体颗粒、液滴或者气体,它们可以分成原生污染和次生污染:原生污染包括SO_x、NO_x和CO,它们主要由自然或工业污染产生;而次生污染则是由原生污染物转化而来,比如地表臭氧层。

烟雾是一种典型的气体污染物,它包括雾和霾。霾是 种由空气中的灰尘、硫酸、硝酸和其他颗粒物组成,并降低能见度的气溶胶系统(主要是PM2.5)。烟雾具有强酸性,危害人类健康和地球生态系统。美国环境保护署指出微小颗粒物的污染会引起严重的健康问题,比如呼吸道疾病、心血管疾病,甚至能致癌和引起生殖发育问题。另外,烟雾不但危害心理健康,还会降低能见度,从而增加交通事故。

烟雾主要由于工业和居民用煤引起。它们主要来自于城市加热,交通排放,和大气中氧化氮和VOCs的光化学反应。空气污染通常集中于人口密集的大型城市,尤其是环境法律法规不健全的发展中国家。[1]例如,1952年12月燃煤引起的伦敦烟雾时间;20世纪四五十年代汽车尾气和工业排放引起的光化学烟雾时间。2013年,中国东北部和东部集中爆发了化石燃料引起的烟雾事件。[2]并且,这场持续很长时间的严重烟雾与地区的气象条件有紧密的关系,当地的亚热带季风气候使得雾霾更加严重。[3]中国于1961—2012年间的雾霾展现了明显的时间和空间特征:雾霾主要发生在冬季,继而才是秋季、春季和夏季(图1-a);中国东部相比西部遭受更多雾霾

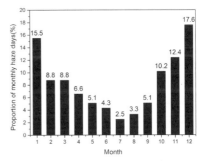

图1-a. 中国中东部地区月雾霾天数(1981—2010年平均)

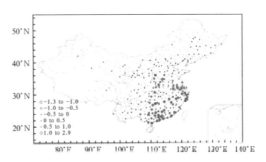

图1-b. 中国冬季雾霾天数线性趋势空间分布(单位:天/年)(1961—2012年)

---

1. A. Ostachuk, P. Evelson, S. Martin, L. Dawidowski, J.S. Yakisich, D.R. Tasat. Age-related lung cell response to urban Buenos Aires air particle soluble fraction [J].Environmental Research107 (2): 170–177, 2008.

2. Green Paper on Climate Change States Causes and Effects of Smog Days(气候变化绿皮书指出雾霾天原因及危害). Southern Weekly (in Chinese). 5 November 2013.

3. R.H. Zhang, Q. Li, R.N. Zhang. Meteorological conditions for the persistent severe fog and haze event over eastern China in January 2013 [J]. Science China, Earth Sciences. 57(1): 26-35, 2014.

设计应对雾霾 DESIGN AGAINST SMOG

图 2. 外滩视角下的 2014 年陆家嘴样貌

图 3-a. 边界层风洞内部

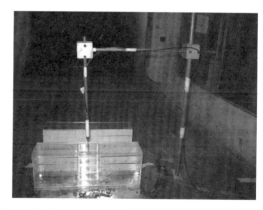

图 3-b. TJ-1 风洞下流方向街道峡谷

天气（图1-b）。[4]

相比WHO健康空气指导线20μg/m³，中国空气污染现状远超国际环境标准。中国严重的雾霾引起了民众对城市地区，尤其是人口密集和高密度楼群的大都市的空气质量的担忧。例如，根据2013年12月1—9日上海地区空气质量指数（AQIs）实时监测数据[5]，9天内PM2.5的平均浓度为211.9μg·m³，达到了重污染水平，其中污染物主要由PM2.5、PM10、NO_2、O_3和CO组成。人口集中区域，如上海陆家嘴、香港、东京和纽约麦哈顿的高密度建筑群导致复杂的城市风环境进而影响气象、空气品质和热岛效应，对城市通风带来了极大的挑战[6]。因此，对城市地区污染物的输运和扩散进行广泛研究非常有必要。

陆家嘴（图2）是位于上海浦东黄浦江东岸的商务区，与旧金融与商业区相对。陆家嘴面积为31.78km²，[7] 从20世纪90年代初作为新的上海金融区得到快速发展。经过20多年的发展，已经树立了数十座超过25层的高层商务大楼。[8] 因此，陆家嘴可作为研究建筑结构与布置对风场和污染物去除过程的代表性地区。

研究方法与应用

经过数十年的发展，理论分析、实验和数值模拟是广泛用来研究风场和污染物扩散的科学方法。

1. 高斯烟雾模型

高斯烟雾模型是最古老也是最流行的方法，主要用来评估连续上升烟气的浓度，这些烟气或污染物来自地表面或者高架污染源。[9] 此模型基于以下几个假设：烟气源自恒定的数学点源，风速和风向、大气湍流在空间和时间上是恒定的。

2. 现场和风洞实验

现场实验是将科学方法应用于实际环境内（或者自然产生的环境），而不是在实验室内进行实验检验[10]，也称之为现场全尺度实验。现场实验优势主要是在实际条件下进行的实验，能提供真实而实际的数据，不会破坏物理上相似性的约束。但这种方法的劣势在于边界条件不可控，不具有重复性，测量点和测量位置有限，并且价钱昂贵、消耗时间。

风洞是在空气动力学领域，采用按比例缩小的模型，研究城市风场和污染物扩散（汽车排放，PM2.5和泄露事故）的重要研究工具，也叫缩尺度风洞实验。同济大学土木工程学院的（环境）大气边界条件（ABL）风洞（TJ1）如图3-a示。从无限空间到有限空间，风洞测试基于三个相似准则来获得可靠的结果，包括几何相似、动力学相似和边界条件相似[11]。为了保证实验模型的准确性，几何原型的无量纲参数必须与风洞实验一致。对于质量守恒方程、动量守恒方程、能量守恒方程和

---

4. L.C. Song, R. Gao, Y. LI, G.F. Wang. Analysis of China's haze days in the winter half-year and the climatic background during 1961-2012 [J]. Advances in Climate Change Research. 5(1): 1-6, 2014.
5. http://www.scmc.gov.cn/aqi/home/Index.aspx.
6. L.N. Yang, Y.G. Li. The influence of building height variability on pollutant dispersion and pedestrian ventilation in idealized high-rise urban areas [J]. Atmospheric Environment. 43: 3111-3121, 2012.
7. https://en.wikipedia.org/wiki/Lujiazui#cite_note-1.
8. Knight Frank China, Shanghai District Overview – Pudong [R].
9. C.H. Bosanquet, J.L. Pearson. The spread of smoke and gases from chimney [J]. Trans. Faraday Soc. 32:1249, 1936.
10. T. van Hooff, B. Blocken. Full-scale measurements of indoor environmental conditions and natural ventilation in a large semi enclosed stadium: Possibilities and limitations for CFD validation [J]. J. Wind Eng. Ind. Aerod. 104: 330-341, 2012.
11. P.Y. Cui, Z. LI, W.Q. Tao. Investigation of Re-Independence of turbulent flow and pollutant dispersion in urban street canyon using numerical wind tunnel (NWT) models [J]. Int. J. Heat Mass Tran. 79: 176-188, 2014.

设计应对雾霾 DESIGN AGAINST SMOG

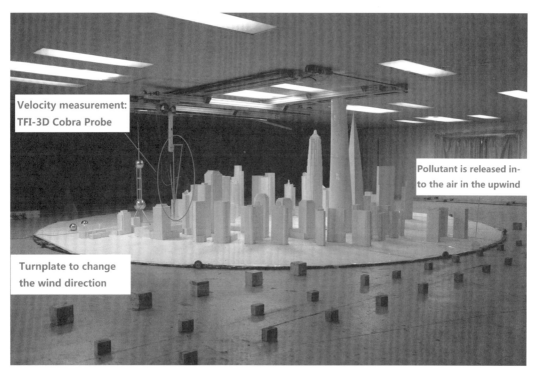

图 4-a. TJ-3 风洞内陆家嘴建筑模型 (1:300)

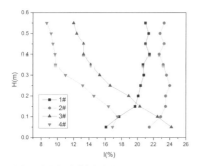

图 4-b. 测量风速随建筑高度变化关系

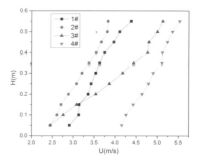

图 4-c. 南风下湍流强度随高度变化关系

图 4-d. 东南风向下,世纪大道释放污染物

浓度守恒方程,Reynolds 数 (Re)、Rossby 数 (Ro)、Peclet 数 (Pe)、Froude 数 (Fr) and Schmidt 数 (Sc) 是通过对控制方程进行无量纲化得到的[12]。在风洞实验中,所有的无量纲参数均满足是不可能的,也是不实际的。Snyder[12,13] 指出对于模拟尺度低于 5km 的原型流动,Rossby 数可以忽略。如果模型处于足够大雷诺数的风场中,Reynolds 数、Peclet 数和 Schmidt 数可以忽略。这样的话,Fr 数是唯一需要满足的无量纲数。对于图 3-b 所示的街道峡谷,建筑临界雷诺数 (ReH,crit) 为 3.3×10^4。[11] 我们采用同济大学的 TJ-3 风洞来模拟陆家嘴地区的流动和污染物扩散 (图 4-a),建筑几何比例为 1:300。测得的风速 (V_z) 和湍流强度 (I_z) 与建筑高度 (z) 的变化关系如图 4-b 和 4-c 所示。通过线性拟合,得到了幂函数关系式:对于风场,函数关系为 $V_z = V_R \cdot (z/zR)^{\alpha}$;对于湍流强度,关系为 $I_z = I_{uR}(z/zR)^{-\alpha-0.05}$,其中 $\alpha = 0.22$,Z_R 和 V_R 是对应的参考高度和流速。图 4-d 表示污染物在稳态时的扩散接近高斯扩散模型,其受风场条件 (风速、方向和湍流度) 和建筑布局等影响。

风洞实验在准确控制边界条件、提高重复性的优势、预测集中建筑对周边环境的影响和实验流场可视化这些方面展现了明显优势。但是,这种方法易于破坏相似性约束,并且能耗大、成本高。

3. 计算流体力学

CFD 是基于计算机模拟的流体力学数值模拟方法,主要涉及流体流动、传质及其如化学反应等伴生现象的分析系统。目前,该方法已被广

---

12. W.H. Snyder. Guidelines for fluid modeling of atmospheric diffusion [R].EPA-600/8-81-009, 1981.
13. W.H. Snyder. Similarity criteria for the application of fluid models to the study of air pollution meteorology [J]. Bound-Lay Meteorol. 3: 113-134, 1972.

设计应对雾霾 DESIGN AGAINST SMOG

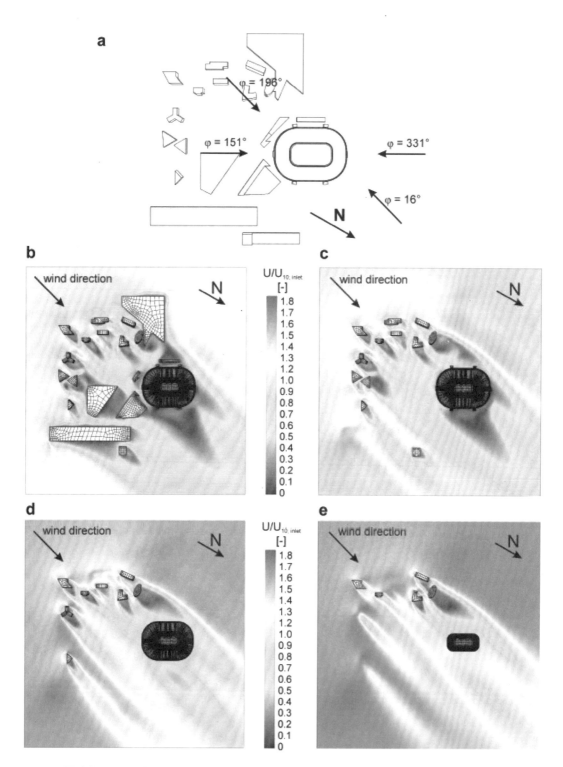

图5-a. CFD模拟中带风向指示的体育场和周围建筑顶视图 (φ = 16°, 151°, 196°and 331°); (b–e) 四个水平面无量纲风速 U/U10 轮廓图 φ = 196° (SSW), U10 = 5 m/s; 图 5-b. 10 m; 图 5-c.20 m; 图 5-d. 40 m and 图 5-e. 60 m

泛应用航空航天、汽车、生物医药、化学、市政环境、动力与体育工程等领域。[14] 与现场实验得到的风场相结合,CFD被用来模拟纽约曼哈顿城市地区的大气流动和污染物扩散。[15] 定量结果显示出:大城市街道内释放的污染物可传播到上风向和侧风向上的几个街区。Van Hooff and Blocken (2010)[16] 采用CFD耦合方法,同时计算了相同计算区域内室外和室内空气流动,例如对Amsterdam ArenA体育馆从2 900m到0.02m尺度范围内分析其自然通风情况。计算结果显示,所有流动变量在计算区域内的室内和室外间直接交换。该方法的优势在于没有采用任何假设来简化计算模型。图5-b到图5-e显示了基于体育馆和周边建筑的风向,在不同高度的CFD模拟结果。ArenA周围较低的风速意味着体育馆位于写字楼的尾部,这个位置会引起风向上较低的通风率。这项研究结果表明微小的几何修正能大幅度增加通风速率,最大值达到43%。

城市尺度范围内空气流动和气体污染物扩散的物理现象实际属于多尺度系统问题。Cui 等人[17] 提出采用多尺度方法,数值研究交通排放对城市尺度范围内室内空气品质的影响。计算区域可按照不同尺度分为:社区尺度、街道尺度和室内尺度。计算从顶端到低端进行,顶端的计算结果在界面处被提取出来,作为边界条件应用于下一尺度的分析。采用这种多尺度数值模拟方法得到的结果与实验结果吻合很好,这意味着多尺度方法对于解决城市地区多尺度问题,如城市到室内,甚至更小尺度间的污染物对流扩散问题具有重要的实际意义。另外,这种方法相比于同时计算整个区域而言更加经济。

相比于分析解和实验方法,CFD方法的最大优势在于其可视化,以及准确的控制物理条件和无需相似准则约束的情况下,得到整个计算区域内流动相关变量的详细信息。但是,如何确保数值计算准确性是CFD工程师最关切的问题。因此,实验验证对数值计算而言必不可少。对城市空气流动的CFD研究需要对实地进行准确的全尺度或者缩小尺度的风洞测量。

减少空气污染策略

中国空气污染主要由工业和汽车中煤和汽油等化石燃料的燃烧引起,因此减小空气污染的有效方法是发展清洁能源,比如风能、太阳能、水能和生物能源,以及通过发展公共交通或者开发电动汽车以减少燃油汽车的使用。短期而言,必须立即采取一些积极措施以控制空气污染对公共健康的影响,比如改善城市规划以加强城市通风和增加绿地;加强公共监督和便于民众获得空气品质信息。为了改善中国空气品质,国务院在2013年9月10日颁布了《大气污染防治行动计划》。该计划涵盖了针对空气污染五年防治后的目标,到2017年的具体指标和实质性的办法。

---

14. W.Q. Tao. Numerical Heat Transfer [M]. Xi'an Jiaotong University Press, Xi'an, 2001.
15. S.R. Hanna, M.J. Brown, F.E. Camelli, et al. Detailed simulations of atmospheric flow and dispersion in urban downtown areas by computational fluid dynamics (CFD) models - an application of five CFD models to Manhattan. Bulletin of the American Meteorological Society, 2006.
16. T. van Hooff, B. Blocken. Coupled urban wind flow and indoor natural ventilation modeling on a high-resolution grid: A case study for the Amsterdam ArenA stadium [J]. Environ. Modell .Softw. 25: 51-65, 2010.
17. P.Y. Cui, Z. Li, W.Q. Tao. Wind tunnel experiments and multiscale modeling for effects of traffic exhausts on the indoor air quality within urban-scale regions [C] // The 13th International Conference on Indoor Air Quality and Climate, 2014, Hong Kong.

Air Pollution Status in Urban Areas and Research Methods

Li Zhuo,
Tao Wenquan

Air pollution is contaminated by any particulates, chemical or biological molecules that modify the natural characteristics of the atmosphere. The pollutants can be solid particles, liquid droplets or gases, which are classified into primary or secondary: the first primary pollutants such as sulfur oxides (SOx), nitrogen oxides (NOx), carbon monoxide (CO) are mainly produced in natural or industrial processes, while secondary pollutants are not from direct emissions but produced by the primary pollutants such as ground level ozone.

Smog is a typical air pollutant that is a combination of fog and haze. Haze is an aerosol system composed of the dust, sulfuric acid, nitric acid and other particles in the air which lead to visual disorder (mainly PM2.5). Smog is highly toxic and has been recognized as threat to human health as well as to the Earth's ecosystems. According to the United States Environmental Protection Agency (EPA) 2004, the fine particle pollution poses serious health threats that can cause respiratory and cardiovascular harm, and might cause cancer, reproductive and developmental harm. Smog also has negative influence on psychological health, and gives rise to the traffic accidents because of the poor visibility.

Smog is mainly caused by the use of coal in industries and individual buildings for heating and transportation emissions in urban areas, and the chemical reaction of sunlight with nitrogen oxides and volatile organic compounds (VOCs) in the atmosphere, and natural environment. Air pollution is usually concentrated in densely populated metropolitan areas, especially in developing countries where environmental regulations are relatively lax or nonexistent.[1] For example, the great smog of December 1952 in London is coal-caused smog, and the modern smog in Los Angeles during 1940s-1950s is due to the vehicular emission and industrial fumes react with sunlight to form light blue photochemical smog. In 2013, a dense wave of smog began in the major cities of northeastern and eastern China due to the increase in fossil fuel consumption.[2] Furthermore, this sustained severe smog event was related to the meteorological conditions that the weakened East Asian winter monsoon exacerbated the occurrence of smog.[3] The haze days in China during the period of 1961-2012 show temporal and spatial characteristics: haze mainly happened in the winter half-year, followed by autumn, spring and summer (picture 1-a); the eastern China suffered much more haze days than the western China (picture 1-b).[4]

Current levels of air pollution in China far exceed international environmental standards, compared to the World Health Organization's healthy air guideline of 20 $\mu g/m^3$. The heavy smog in China has raised concerns on the air quality in urban areas, especially in the municipalities with large populations and dense building groups. For example, based on the real-time monitoring data of Air Quality Index (AOIs) during 1st to 9th December, 2013 in Shanghai,[5] the averaged concentration of PM2.5 over 9 days is 211.9 $\mu g \cdot m^3$, which belongs to the heavily polluted level, and the pollutants are mainly composed of PM2.5, PM10, NO_2, O_3 and CO. The high density building groups in densely-populated urban areas such as Lujiazui in Shanghai, Hong Kong, Tokyo, and Manhattan in New York have posed challenges to urban ventilation, by generating complex city wind environment which involves meteorology, air quality and heat island effect.[6] Therefore, it is necessary for comprehensive studies to understand the transport and dispersion characteristics of such pollutants in urban areas.

Lujiazui (picture 2) is a central business district (CBD) located in Shanghai, on the eastern bank of the Huangpu River in Pudong district and sits opposite to the old financial and business district of the Bund. Lujiazui area is 31.78 km² (12.27 sq mi) [7] and has been developed specifically as a new financial district of Shanghai since the early 1990s. After more than twenty years developments, there are more than 30 high-rise buildings over 25 stories for commerce. [8] Lujiazui is a very representative area to understand how the structures and layout of the buildings affect the wind flow and the pollutant removal process.

Research Methods and Applications

Over the last decades, three complementary research methods are most commonly used to study the air flow and pollutants dispersion which are theoretical analysis, experiments and numerical simulation.

1. Gaussian plume model

Gaussian plume model is the oldest and perhaps the most popular one to analytically assess pollutant concentrations of continuous, buoyant air pollution plumes originating from ground-level or elevated sources.[9] This model is based on several assumptions that the plume starts from a constant mathematical point source, the wind speed and direction, and atmospheric turbulence are constant in space and time.

2. Field and wind tunnel experiments

A field experiment applies the scientific method to experimentally examine an intervention in the real world (or naturally occurring environments) rather than in the lab, [10] also called on-site full-scale experiment. The advantages of field experiments include that it is conducted on the real conditions and provides the real and realistic data, and doesn't violate similarity constraints in physics. The disadvantages of this method are that it suffers from uncontrollable boundary conditions and lacks of repeatability, the measurement points and its positions are limited, and it is very expensive and time consuming.

A wind-tunnel is an important tool in aerodynamic research to study the air flow and pollutant dispersion (vehicle exhaust, PM2.5 and accidental leakage) in urban areas with reduced-scale models, also called reduced-scale wind tunnel experiment. Building (Environmental) Atmospheric Boundary Layer (ABL) wind tunnel (TJ1) in Tongji University is shown in picture 3-a. From the infinite space to finite space, wind tunnel test with reduced-scale models should follow three similarity principles to obtain the reliable results for the real design, which are geometric similarity, kinematic similarity and boundary condition similarity.[11] For an accurate modeling, several dimensionless parameters in the prototype must be duplicated in the wind-tunnel experiments. In terms of the conservation equations of mass, momentum, energy and concentration, five dimensionless parameters of Reynolds (Re), Rossby (Ro), Peclet (Pe), Froude (Fr) and Schmidt (Sc) numbers were obtained through nondimensionalization of the governing equations.[12] In wind-tunnel experiments, the duplication of all the dimensionless parameters are impossible and impractical. Snyder [12, 13] pointed out that the Ross by number can be neglected when modeling prototype flows with a length scale less than about 5km. The Reynolds number, Peclet number, and Schmidt number criteria may be neglected if the model flow is sufficiently high Re. Then in such a case, Fr is the only one which should be matched between the prototype and the wind-tunnel tests. For the street canyon studied shown in picture 3-b, such determined critical building Reynolds number (ReH,crit) is 3.3×10^4.[11] TJ-3 wind tunnel in Tongji University was used to model the flow and pollutant dispersion within Lujiazui region with the ratio 1:300 as presented in picture 4-a. The measured wind velocity (Vz) and turbulence intensity (Iz) varying against the building height (z) are given in picture 4-b and 4-c, respectively. By fitting these curves, the following power law functions are obtained: for the wind flow, the function is given as $V_z = V_R \cdot (z/zR)^\alpha$; for the turbulence intensity, it is $I_z = I_{uR}(z/zR)^{-\alpha-0.05}$, where $\alpha = 0.22$, zR and VR are corresponding to the reference height and velocity. picture 4-e shows the pollutant dispersionat steady state is close to Gaussian diffusion model, influenced by the wind conditions (wind speed, directions and turbulent intensity), building configuration, etc.

Wind-tunnel experiments provide advantages of accurately controlling boundary conditions and high repeatability, predicting the intended building effects on surroundings, and realizing flow field visualization. However, this method tends to violate similarity constraints, and is high cost and energy consuming.

3. Computational fluid dynamics (CFD)

CFD is the most common numerical method, which is an analysis system involving fluid flow, heat transfer and associated phenomena such as chemical reactions by means of computer-based simulation. This method has been widely applied in aerospace, automotive, biomedical, chemical, civil and environmental, power and sports engineerings, et al [14]. Combined with wind flow observations obtained by field experiment, CFD was applied to simulate the atmospheric flow and dispersion in urban areas of Manhattan, New York [15]. The quantitative results showed that transport of a release at street level in a large city could reach a few blocks in the upwind and crosswind directions. Van Hooff and Blocken (2010)[16] conducted coupled CFD approach to simultaneously solve the outdoor and indoor air flow within the same computational domain, for examples analyzing natural ventilation for the geometrically complex Amsterdam ArenA stadium with length scales ranging from 2,900m to 0.02m. The flow through the ventilation openings was solved explicitly and all flow variables were directly transferred between the outdoor and indoor parts of the domain. The advantage of their method is that no assumptions are premised. Based on the wind direction around the stadium and surrounding buildings, the CFD results of the wind flow field at different heights φ = 196° (SSW) are presented in picture 5-b to 5-e. The lower velocity values around the ArenA indicate that the stadium is indeed situated in the wake of the office buildings, causing the lower air exchange rate for this wind direction. Their results show that small geometrical modifications result in increasing the ventilation rate by up to 43%.

The physical phenomena of air flow and dispersion of gaseous pollutants in urban-scale regions actually belong to the multiscale system problems. Cui et al.[17] proposed a multicale approach to numerically study the impacts of traffic exhausts on the indoor air quality in urban-scale regions. The computational domain is divided into three regions with different scale, which are community level, street level and indoor level. The computational processes start from top to down level and the computational results from the top level are extracted at the interface to be applied as the boundary conditions on the next level analysis. The numerical results agree well with the experimental results that indicate the multiscale method is practical to solve the multiscale problems in the urban areas, such as the pollutant flow and dispersion from urban level to indoor level, even smaller regions. Furthermore, this method is economically compared to using the method by solving the whole domain simultaneously.

Compared to the analytical solution and experimental methods, the main advantages of CFD are obvious that can provide detailed visualization and the information on the relevant flow variablesin the whole calculation domain under well-controlled conditions and without similarity constraints. However, how to guarantee the accuracy of numerical results is the primary issue for CFD engineers. Experimental validations thus are imperative for numerical simulations. For CFD studies of urban wind flowrequires high-quality on site full-scale or reduced-scale wind tunnel measurements to be compared with the simulation results.

Efforts to Reduce Air Pollution

Air pollution in China is mainly caused by combustion of fossil fuels in industries and vehicles such as coal and oil, thus an effective means to cut down air pollution is to develop clean power sources such as wind power, solar power, hydropower, bioenergy and reduce the use of cars by developing the public transport or conversing to electric vehicles. In the short term, the passive means to control the impact of air pollution on public health have to be taken action immediately: improving urban planning to strengthen city ventilation and increasing green spaces; strengthen public supervision and making the public easily obtain complete air quality information. In order to improve air quality in China, the State Council of the People's Republic of China issued "The Action Plan for the Control of Air Pollution" on September 10, 2013. This plan includes the goal after five years of efforts, specific indicators by 2017, and substance methods to fight against air pollution.

设计
成果

WORKS

设计应对雾霾 DESIGN AGAINST SMOG

雾霾影响着成千上万人的生活，这是一个城市层面的问题。我们试图以此为契机，重新规划陆家嘴，改善陆家嘴区域的空气状况与生活质量。热动力学的核心议题，即关注空气流动、热辐射、湿度等与人体舒适度相关的基本物理元素。我们试图以热力学的设计方法重新定义陆家嘴，增加热力学基础设施，提高多功能建筑物的密度，减少汽车使用，使之从利于车行变为利于人行的区域，从工作之城变为生活之城。

Smog affects thousands of people, apparently it's a problem on urban level. Our proposal embraces Lujiazui with its successes and failures and re-plan it to improve the air condition as well as life quality. Thermodynamic theories emphasize on physical elements such as air flow, thermal radiance and humidity related to human comfort, which is exactly the core of this topic. Therefore, a new masterplan of Lujiazui is designed thermodynamically to decrease car usage and increase density on infrastructure and mixed-use buildings, making it from car-friendly to pedestrian-friendly, from a working city to a living city.

deMONSTERative Thermodynamics
非"异形"热力学

Student 学生：
Caio Barboza 卡约·巴博萨
Christian Lavista 克里斯蒂安·拉维斯特
Ramus Guillaume 拉姆斯·支纪尧姆
Louis Jean 路易斯·让
Wang Liyang 王立杨
Yang Zhiyun 杨之赟
Qian Ren 钱牣

Supervisors 指导：
Li Linxue 李麟学
Marta Pozo 马韬

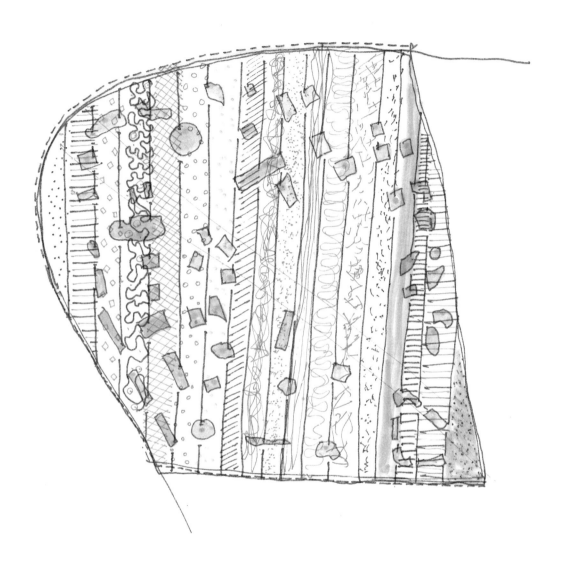

早期细分草图 1
Early Sketches on Detailed Programs 1

设计应对雾霾 DESIGN AGAINST SMOG

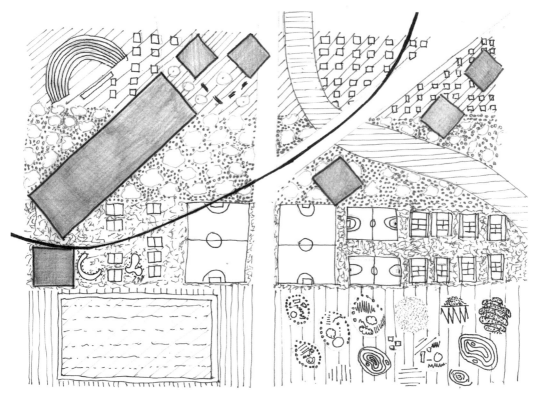

早期概念草图 2
Early Sketches on Detailed Programs 2

我们讨论过多种应对雾霾的策略，比如用具有传播性的小尺度元件吸收雾霾，由流浪汉、清洁工、汽车等可移动的物体携带；再如利用飞机起飞降落穿越大气层的特性作为接触媒介减少雾霾的产生；或者用某种方法将雾霾储存在建筑体中，夏季反射太阳，冬季释放热量。然而与这些想法相比，我们认为能够从源头上减少雾霾、改变人们生活方式的是提升规划。既然陆家嘴区域存在着各种各样的问题，为何不将 2015 年的理解回溯至 1987 年？

We discussed a lot on strategies against smog. It can be some small device absorbing pollution to be dispersed by movable objects such as homeless people, cleaners, cars. Another is to use airplane as a medium to decrease smog level, taking advantage of its contact with air when taking off and landing. Besides, in some way smog can be stored in architecture, reflecting sunlight in summer and releasing heat in winter. However, comparing to these ideas, it is a new masterplan that can reduce smog from source and change people's life style. Since there are kinds of problems in Lujiazui district, why don't we go back to 1987 with 2015's understanding?

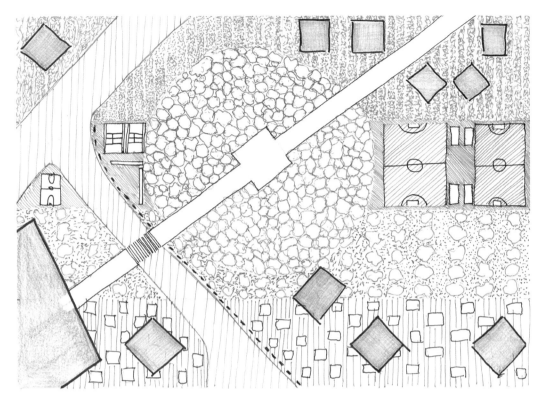

早期概念草图 3
Early Sketches on Detailed Programs 3

最初的想法是将汽车道路去除，消解街区的分界，将平行的活动带铺满基地。可能会产生强烈的戏剧性场景，比如摩天高楼出口对面是一群人在打麻将。这种对立在我们看来有利于激活城市活力，避免单一城市的出现。

The first idea is to pave it with programmatic bands by removing car avenues and dissolving block boundaries. A possible dramatic scene is that people are playing mahjong on the opposite of a skyscraper exit, which is a way to activate the city instead of simple city common in CBD area.

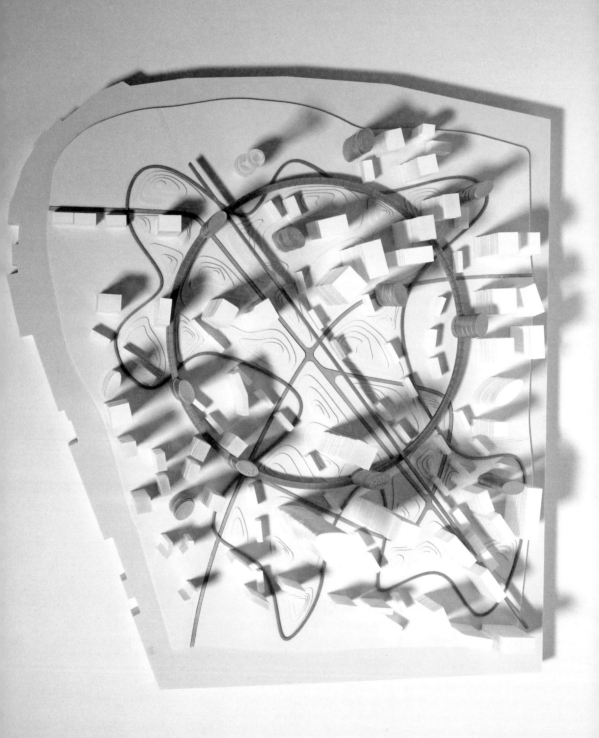

非"异形"热力学 deMONSTERative Thermodynamics

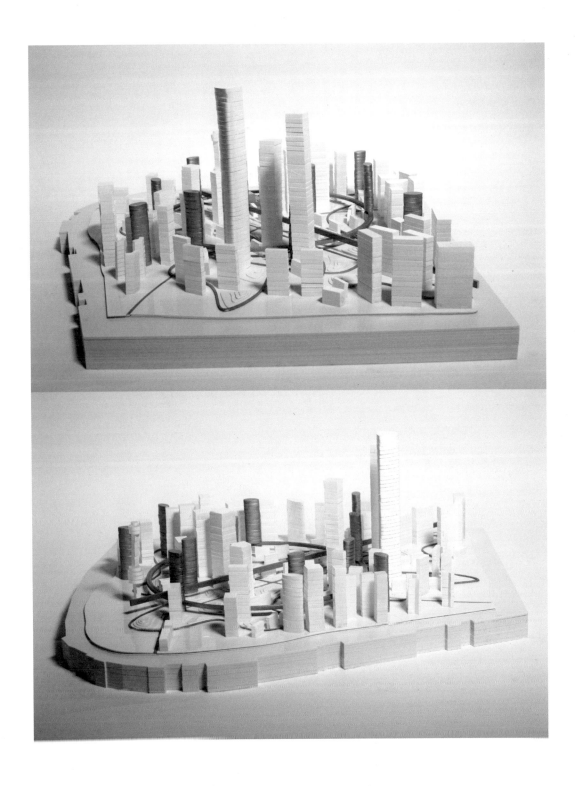

设计应对雾霾 DESIGN AGAINST SMOG

16 % - 36 %
NEIGHBOURING PROVINCES

64 % - 84 %
SHANGHAI

JIANGSU

ZHEJIANG

区域性雾霾影响 Regional Smog impact

随着中国城市化进程的推进以及私人汽车的不断增加，汽车在可见的未来还会持续增多。从上海雾霾构成图表来看，大部分雾霾由上海自身产生，其中最多的是汽车在生产和日常使用过程中产生的有害气体，与此同时过大的道路尺度产生了很多城市问题。因此，减少汽车使用是本提案的核心。陆家嘴处于浦东和浦西的连接中心，需要更多的基础设施连接两岸。高效的交通系统作为汽车的替代品置入基地中，连接黄浦江的东西两侧。

According to the potential of growing urbanization of China and the sustained increasing private cars, the car use will still be increased in the foreseeable future. Per the chart of Regional Smog Breakdown, the majority of smog is produced by Shanghai itself and cars contribute most in both producing and consuming way, which also results in urban problems in Lujiazui district. Thus in our proposal, the decrease of car usage is the key to control smog. As a middle point of access between Pudong and Puxi, more infrastructure are needed in Lujiazui District to connect both sides. In consequence, we introduce a hyperloop into site as an efficient transportation to replace cars, linking both sides of Huangpu River.

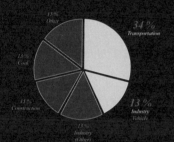

上海雾霾构成 Regional Smog Breakdown

34 % Transportation
13 % Other
13 % Coal
13 % Construction
13 % Industry (Other)
13 % Industry Vehicle

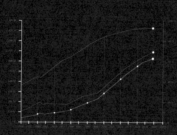

城市化进程 Urbanization in China

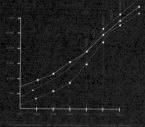

汽车经济 Car Economy

基地与浦东/浦西的可达性 Accessibility Pudong/Puxi

利用 CFD 软件 Phoenix 测算出人行高度夏季与冬季风的分布情况，以此确定风的流向以及湍流区域。在湍流区域增加通风塔，改善通风情况。同时沿着主导风向引入 3~5 层的低层建筑，在增加建筑密度的同时强化风的流向。

With the help of CFD software Phoenix, major wind contribution as well as the turbulence in Lujiazui district can be identified. Thermodynamic towers are added in turbulence point to improve air flow here and 3~5 store buildings are introduced along the direction of predominant wind flow to not only increase the density but also reinforce the ventilation.

非"异形"热力学 deMONSTERative Thermodynamics

夏季风（设计前）Summer - Wind - Before

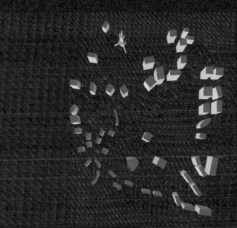

冬季风（设计前）Winter - Wind - Before

夏季主导风向与湍流区域 Summer - Predominant Winds and Turbulence

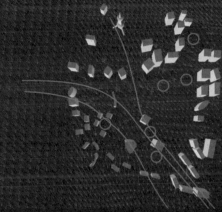

冬季主导风向与湍流区域 Winter - Predominant Winds and Turbulence

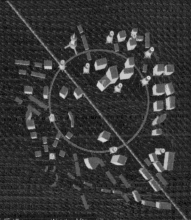

夏季风（设计后）Summer - Wind - After

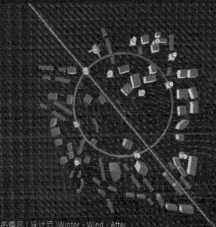

冬季风（设计后）Winter - Wind - After

设计应对雾霾 DESIGN AGAINST SMOG

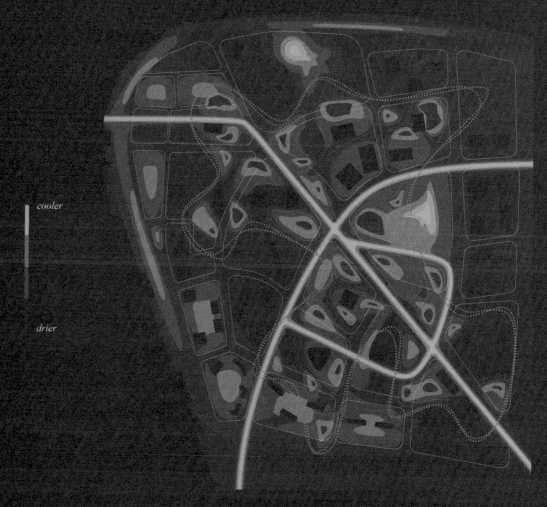

cooler

drier

基于热舒适度的人行道路及水系
Pedestrian Path and Water Body Added - Based on Thermal Comfort

A　　　　　　　　　B　　　　　　　　　C

设计生成步骤
Generating Steps

非"异形"热力学 deMONSTERative Thermodynamics

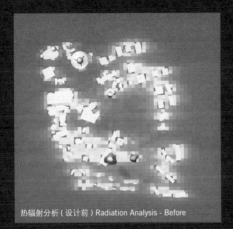

热辐射分析（设计前）Radiation Analysis - Before

热辐射分析（设计后）Radiation Analysis - After

服务空间	Service	0%
文化空间	Cultural space	0%
办公空间	Office	85%
居住空间	Residential space	5%
商业空间	Commercial space	10%

服务空间	Service	5%
文化空间	Cultural space	5%
办公空间	Office	65%
居住空间	Residential space	10%
商业空间	Commercial space	15%

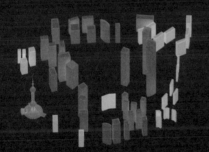

在区域湿度图的基础上，得到了基于人体舒适度的人行道路，并根据 1935 年的历史地图将陆家嘴区域的水体恢复。热辐射图的模拟图则展示了夏季设计前后的温度变化，人行道路内部的植被景观提供更多的遮挡，此区域的热舒适度得以提升。

Based on the humidity map, we trace the pedestrian path on human comfort and rebuild the water body referring to the Shanghai map of 1935 to provide more humidity. A simulation is performed to compare the before-after impact on temperature in summer. With vegetation providing more shades, the area in the pedestrian path has a better thermal comfort.

设计应对雾霾 DESIGN AGAINST SMOG

陆家嘴区域总平面
Masterplan of Lujiazui District

高速轨道连接着浦东与浦西，内部的环路轨道连接着作为环路站点的风塔和既存的塔楼。从而使得此区域内人们通过步行即可到达任何位置。我们将汽车道路改造成为开放的植被景观与户外活动设施，以此创造一个富有活力的社区。

Hyperloop links Pudong and Puxi while inner loop connects each wind towers as stations and existing towers, for this reason anywhere can be accessible on walk in this area. Car avenues are renovated into open landscape full of vegetation as well as outdoor facilities, thus creating an active neighborhood.

非"异形"热力学 deMONSTERative Thermodynamics

拼贴展示了陆家嘴区域未来的图景,汽车从城市中消失不见,取而代之的是布满植被与各类活动设施的步行空间。并此,自然景观、回溯性的河流、漂浮的基础设施、风塔和已有的塔楼混合在一起,形成未来都市的一种新范式。

The collage shows the future image of Lujiazui district. Cars are replaced by a pedestrian space covered by all kinds of vegetation and facilities. Here, natural landscape, retroactive water system, upper-level infrastructure, mixed-used wind tower and existing towers are mixed together as a new paradigm of future urbanism.

设计应对雾霾 DESIGN AGAINST SMOG

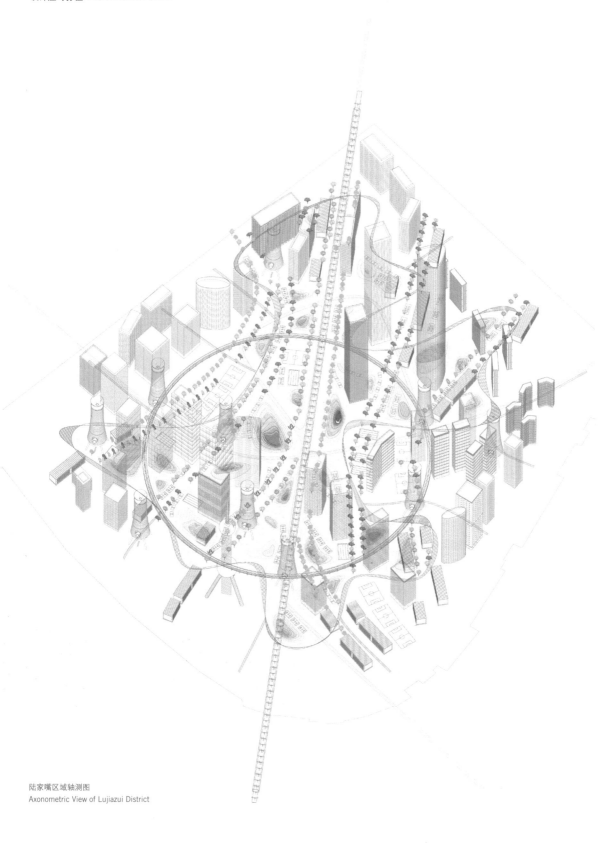

陆家嘴区域轴测图
Axonometric View of Lujiazui District

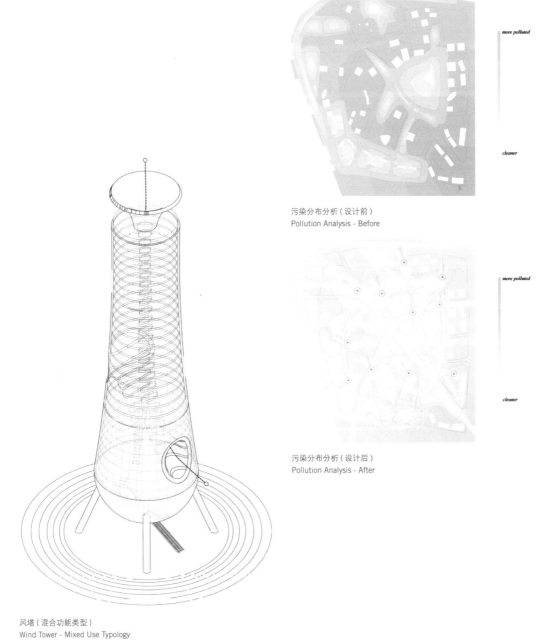

污染分布分析（设计前）
Pollution Analysis - Before

污染分布分析（设计后）
Pollution Analysis - After

风塔（混合功能类型）
Wind Tower - Mixed Use Typology

风塔在冬夏分别利用风压和热压通风的机制改善风环境，以此破坏雾霾的形成条件。风塔作为环路的站点，还混合着商业、住宅和办公等其他功能，帮助提升陆家嘴区域的建筑密度。

Wind tower improves the wind environment to prevent the formation of smog. It serves both season utilizing wind pressure ventilation in winter and thermal ventilation in summer. Meanwhile, as the stations of inner loop, wind tower is a mixed use typology including commercial, residential and office, which helps to increase the density as well.

设计应对雾霾 DESIGN AGAINST SMOG

cooler ▬▬▬▬▬▬▬▬ drier

cleaner ▬▬▬▬▬▬▬▬▬▬▬▬▬▬ more polluted

湿度及污染分析（设计后）
Overlayed Humidity and Pollution Situation - After

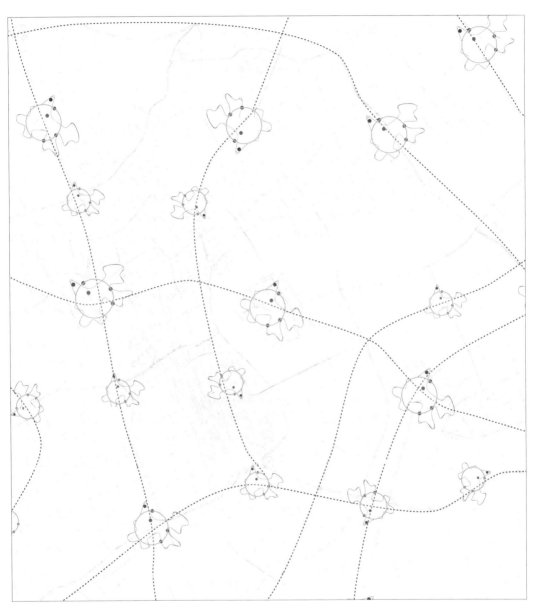

陆家嘴范式的应用潜力
Potential Application of Lujiazui Paradigm

我们设想陆家嘴区域可以成为一种城市中的范式，而后推广到上海或者中国的其他区域，从而真正意义上改变每个人的生活。

Our proposal of Lujiazui district can be an urban paradigm, which is to be applied to other area in Shanghai or the rest of China and change everyone's life style in a real sense.

设计应对雾霾 DESIGN AGAINST SMOG

我们设计的不是一座建筑，而是一种建筑设计的策略。我们试图研究为什么现在的设计会制造雾霾，而我们对此又能有何作为。我们不想采取过激的行动加强人们对雾霾问题的认识，也不想只是隔靴搔痒，暂时缓解城市之痛。我们想探索可以更好的利用最基本的热力学定律、建造真正减少污染物排放的基础架构。

Our proposal is not a single finished building, but a detailed architectural strategy for design. We look at the reasons why our infrastructure produces smog, and what we can do. We are not interested in simply making a provocation that makes people aware of smog. We are not interested in a Band-Aid solutions or painkillers, that helps only right now. We are interested in searching for ways that the most fundamental thermodynamic principles can be leveraged to imagine a holistic infrastructure, which releases less pollution.

Invisible Infrasturcture
不可见的基础设施

Student 学生：
Alexander Matthias Jacobson 亚历山大·马蒂亚斯·雅各布森
Inés Brotons Borrell 艾尼斯·布洛东·博雷尔
James Arthur Clive Hargrave 詹姆斯·亚瑟·克莱夫·哈格雷福斯
Zhang Xiang 张　翔
Zhang Runze 张润泽
Zhu Jingyi 朱静宜

Supervisors 指导：
Zhou Jianjia 周渐佳
Gao Jun 高军

不可见的基础设施 Invisible Infrastructure

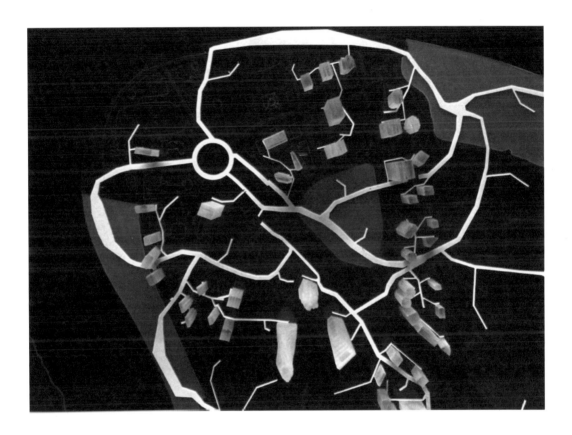

模型照片
3D Model

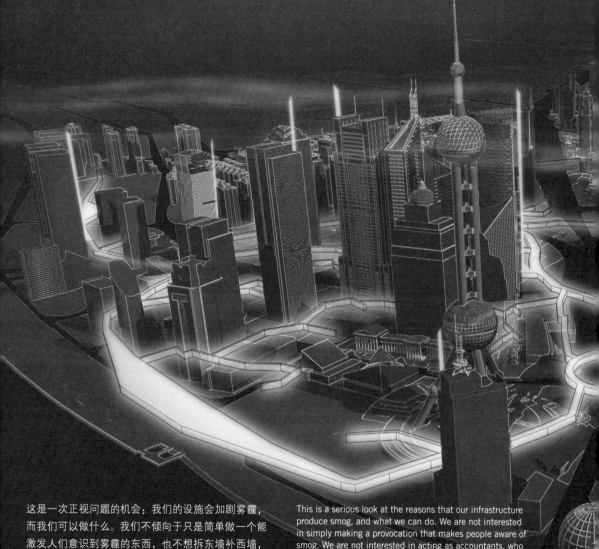

这是一次正视问题的机会：我们的设施会加剧雾霾，而我们可以做什么。我们不倾向于只是简单做一个能激发人们意识到雾霾的东西，也不想拆东墙补西墙，把雾霾从一个地方搬到另一个地方，因为我们并不需要一个创可贴式的解决方案——仅仅为了当下的缓解。我们立志找到基本热力学原则下的全局观控制，并真正帮助解决空气污染问题。我们提出了一种新的可以融入全局设施中的建筑布局。

This is a serious look at the reasons that our infrastructure produce smog, and what we can do. We are not interested in simply making a provocation that makes people aware of smog. We are not interested in acting as accountants, who shift the production of smog from one place to another. We are not interested in a band-aid solutions or pain-killers, that help only right now. We are interested in searching for ways that fundamental thermodynamic principles can be leveraged to imagine a holistic infrastructure which releases less pollution. We propose architectural configurations that participate in this holistic infrastructure.

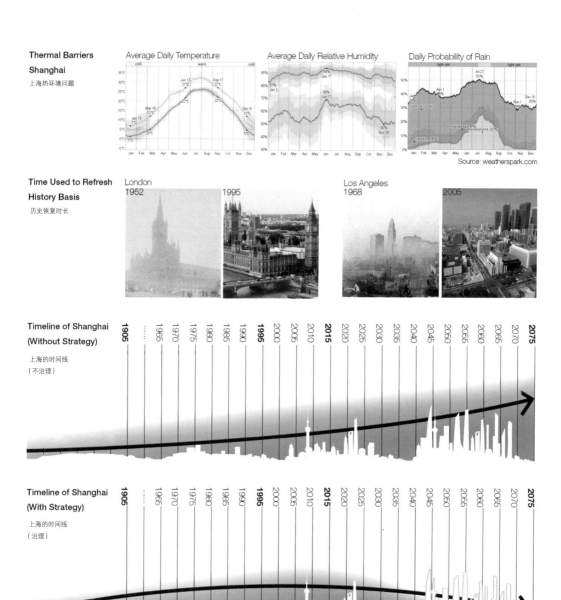

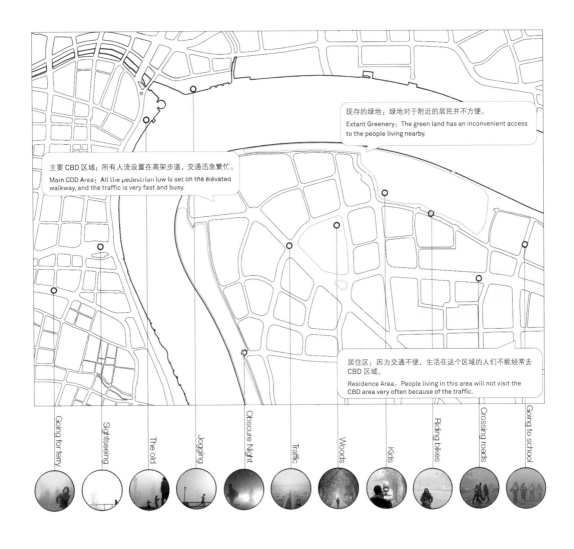

作为建筑师，我们关注城市的多个层面，雾霾便是其中之一。回顾伦敦和洛杉矶两座城市，雾霾是一个仅有60年历史的慢性问题，但是城市介入和产权结构的历史更长。因此我们必须从整体层面思考城市生活的长期质量。

陆家嘴区域的三个主要问题：
1. 道路在城市尺度上扮演了壁垒的角色。
2. 人们不得不每日穿梭数公里从家里赶到公司上班。
3. 汽车的过度使用，产生了大量雾霾。

As architects, we have many concerns of the city, and smog is one of these concerns. If we look at the example of London or Los Angeles, smog was only a chronic problem for 60 years, but urban interventions and ownership structures last much longer. So we must think holistically to address the long-term quality of life in the city.

Three Main Problems in Lujiazui district:
1. At the urban scale, the roads function as barriers.
2. The people have to travel several kilometers between home and office.
3. Excessive use of cars, which produces smog.

设计应对雾霾 DESIGN AGAINST SMOG

此处已经存在一个正在继续扩建的高架人行系统,基于此我们也许可以赋予人行系统新的意义。

There is already precedent for constructing such pedestrian walkways, we propose a more sophisticated understanding of the possibilities that these pedestrian paths might offer.

这些通道用来连接陆家嘴地区的建筑,所以整个城市的功能都被这个热力学引擎连成一个系统

The pathways themselves can be used to interconnect the buildings of Lujiazui, so that the entire city functions as a system of interconnected thermal engines

不可见的基础设施 Invisible Infrasturcture

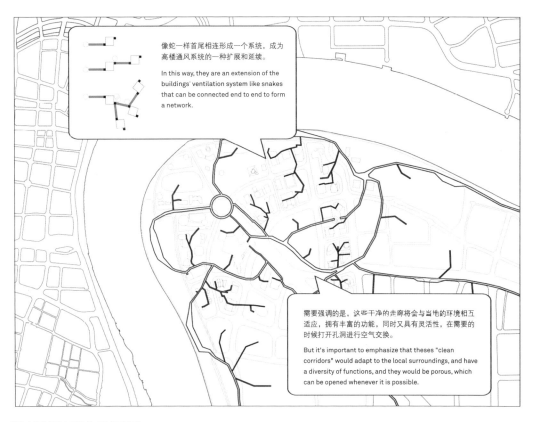

类似血管的网络与现存的建筑相互关联
This vascular network would connect existing buildings to one another

设计应对雾霾 DESIGN AGAINST SMOG

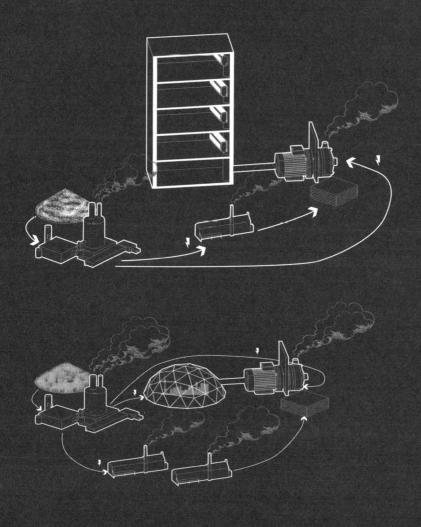

我们首先研究了基地陆家嘴地区存在的问题。从城市尺度来说,这里的道路分布是一大障碍。市民工作与居住之间往往有几千米的距离,这就导致了大量小汽车的使用,同时也制造了雾霾。因为人们没有其他的通行方式,所以是在不自知的情况下助长了雾霾的产生,即使他们选择清洁的出行方式,也同样要受到别人排出的雾霾的影响。

We look at the problems that are present in Lujiazui, our site. At the urban scale, the roads function as barriers. And the people have to travel several kilometers between home and office. This results in excessive use of cars, which produces smog. As a result, individuals involuntarily produce smog because they don't have alternative transport options, and they are exposed to this smog even if they use alternative options.

不可见的基础设施 Invisible Infrasturcture

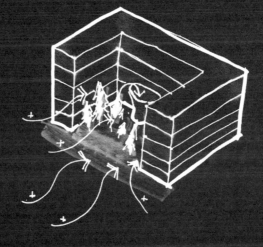

同时我们也仔细研究了基地上可供利用的众多资源。陆家嘴被黄浦江环绕，江水可以被用来进行空气的预热与预冷。这一地区有公园和绿地，可以使用其中的植物对空气进行净化。陆家嘴地区以众多的摩大楼而闻名，这些摩天楼可以用来做热压烟囱，实现空气的自然流动。同时这一地区还有已经初具雏形的高架人行步道系统，我们希望可以将其进一步扩展，延伸到整个区域。这一系统可以被用来将整个区域的建筑联系在一起，形成整体的热力学系统。

We also look at the resources we get from the site. Lujiazui is surrounded by Huangpu River. The river provides ideal cooling and heating. We have parks and green spaces that can act as purifier. Lujiazui is famous for its abundance of high-scrapers that have the potential of acting as solar chimneys. There is also the existing elevated walkway that can be further extended. We propose to extend the existing network of elevated pedestrian paths. The pathways themselves can be used to interconnect the buildings of Lujiazui so that the entire city functions as a system of interconnected thermal engines.

设计应对雾霾 DESIGN AGAINST SMOG

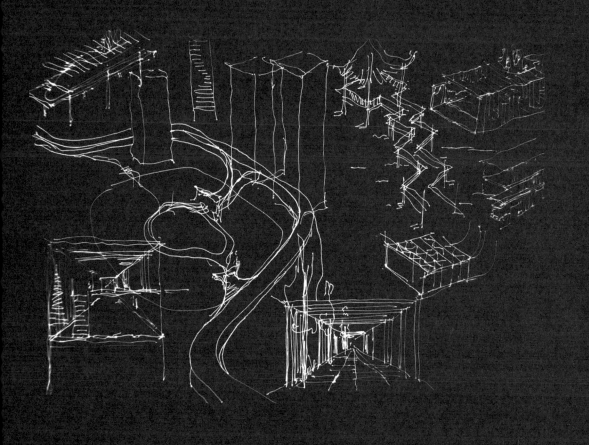

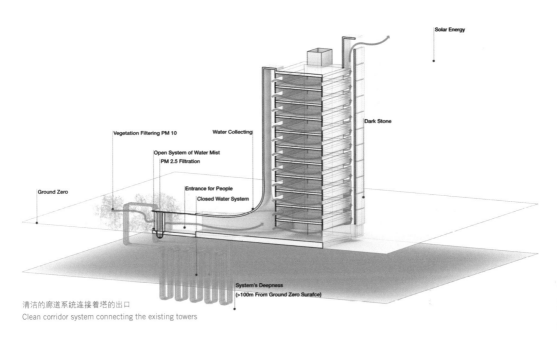

清洁的廊道系统连接着塔的出口
Clean corridor system connecting the existing towers

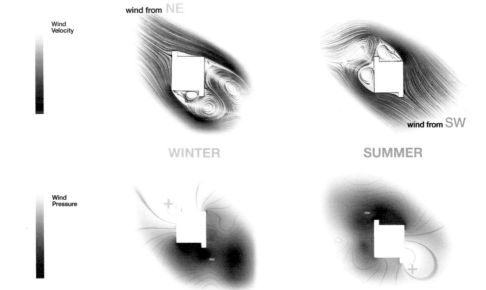

陆家嘴风流模拟
Lujiazui Wind Flow Simulation

因此，我们的提案是一套连接了所有高层的净化走廊系统。走廊的一端是空气的进入口，一般而言放在绿地之中，另一端与作为热力烟囱的高层建筑相连。走廊在延伸的过程中有不同的形态，以适应具体的环境条件与功能需要。我们希望这个走廊是多功能的，希望市民从中获得健康的生活方式，而不仅仅是一次选择的机会。

Thus, we propose a clean corridor system connecting the existing towers. One end of the corridor is connected with an air inlet, typically located in a green space, and the other end is connected to a high-rise building that acts as a thermal chimney. The corridor adopts different configurations along the way, adapting to the surrounding onvironment. We want this corridor to be multi-functional, giving the citizens possibilities of healthier lifestyle instead of a one-time choice.

设计应对雾霾 DESIGN AGAINST SMOG

本设计从雾霾的产生和传播入手，将解决雾霾的重点放在了高层林立的陆家嘴的上海三大高层中央，也就是雾霾容易集聚且不易驱散的空间。在设计之初，通过对雾霾的产生和传播分析得出以下结论，雾霾的产生和传播途径有很多，根治雾霾需要多种手段并存。但如果从空间和环境角度入手，最为有效的则是采用寄生建筑的方法设计小型装置依附在现有高层表皮上，同时根据不同需求创造结构形式，将其应用在新的建筑和现有城市结构上。

此外，为解决过度消耗和浪费问题，并且为将来使用提供最优空间，我们设计了多种不同结构依附于现有建筑上，最小程度干扰原有建筑的结构和空间，为此我们联想到利用帐篷结构创造不同使用功能的私密空间，从而满足人们对不同空间的需求，如花园、阳台等。在结构形式的选择上，我们借鉴了仿生设计的手段。借鉴蝴蝶破茧而出的过程，使展开面积达到最大，结构在高空时更为稳固，吸收阳光并为自身所用。

Our design originates from the generation and dissemination of smog, thus started by focusing on the cener of three high-rises in Lujiazui, Shanghai, where smog is easy to assemble and hard to disperse. Early in the design, through the generation analysis of smog including the generation and transmission of smog, we found that there should be a variety of means to cure smog. But from the space and environmental perspective, the most effective way is through the use of parasitic architecture, to design small devices attaching to the existing epidermis. Besides, we also created verified structures according to different needs, which can be used in new buildings and existing urban structure.

In addition, in order to solve the problem of excessive consumption and waste to provide optimal space for future use, we designed a variety of different structures attached to existing buildings, with minimal disturbance of the original structure of the building and space, and we think that the use of tent structure to create different private function spaces to meet the verified demands, such as gardens, balconies etc. In the choosing of structure, we referred to the biomimetic design. Referring to the process of a butterfly coming out of a cocoon, we finally decided to use a more stable structure at high altitude to absorb sunlight for their own use.

Purify
Ecological Intervention
净化：生态介入

Student 学生：
Adeline Conrad 艾德琳·康拉德
Giulia Esopi 茱莉亚·埃索皮
Du Jie 杜 杰
Ming Lei 明 磊
Zhang Bohan 张博涵
Wu Xiao 吴 潇

Supervisors 指导：
Tan Zheng 谭峥
Li Zhuo 李卓

在对陆家嘴地区的前期研究中，通过计算机模拟，可以看出三个高层中央区域是空气中有害物质易集聚的区域，且高层的背风面更容易附着空气中有害物质。

Through wind simulation of Lujiazui area, results show that the center of three high-rises is the place pollutants gather. What is more, the leeward side of three high-rises are easier to gather pollutants.

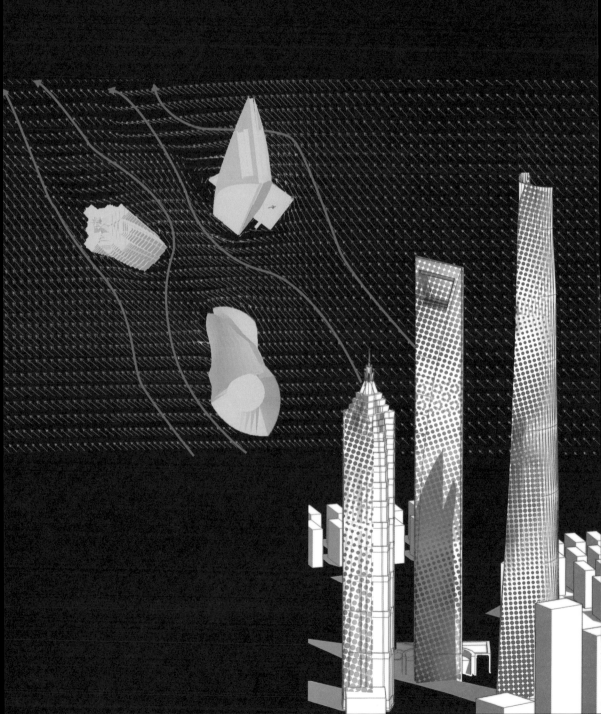

净化：生态介入 Puify: Ecological Intervention

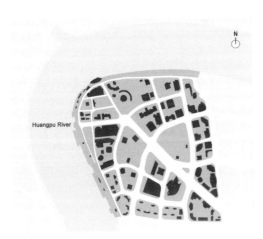

雾霾分布
Smog Distribution

玻璃幕墙分布
Curtain Distribution

风分布
Wind Distribution

绿化分布
Green Distribution

温度分布
Temperature Distribution

设计应对雾霾　DESIGN AGAINST SMOG

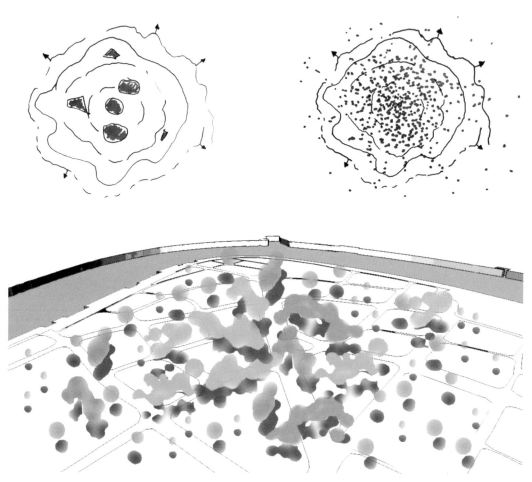

现有绿化分布 vs 未来绿化分布
Existing Green Distribution vs Future Green Distribution

净化：生态介入 Puify: Ecological Intervention

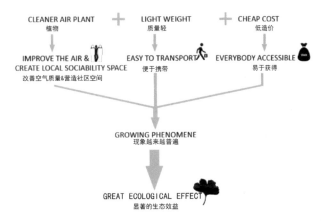

屋顶 On The Roof
立面 On The Facade
地面 On The Ground

传统净化空气的方式依赖于空调以及植物绿化，他们可以在特定的空间、环境中发挥作用。在我们对上海城市绿化分布的研究中发现，在城市层面上，上海市的绿化相对集中，而污染物相对分散，要在城市层面处理污染问题，传统的方法就不太有效了。一种可移动式的装置可以解决这种问题，该装置体积要很小，便于携带和移动，可以闭合与打开，折叠和伸展。该装置相当于一个可移动的绿化设备，可以依附在高层建筑的表皮、屋顶、阳台等部位，也可以安装在地面上。

The traditional way to clean air is using air-condition or relying on green plants inside the city. They could solve the problem in a certain space, refine the air condition where they are arranged. And also in our survey, we find that the green space in shanghai is relatively centralized. In terms of solving the environment pollution in urban scale, traditional methods are not very effective. Pollutants like PM2.5 are in dispersal condition, so a device which is mobile is needed. It should be small enough to be taken anywhere, and it could be folded and stretched, closed and open. And we consume that this device could attach on the façade and roofs of high-rises. And it could also attach on the ground.

设计应对雾霾 DESIGN AGAINST SMOG

结构设计
Structure Design

净化：生态介入 Puify: Ecological Intervention

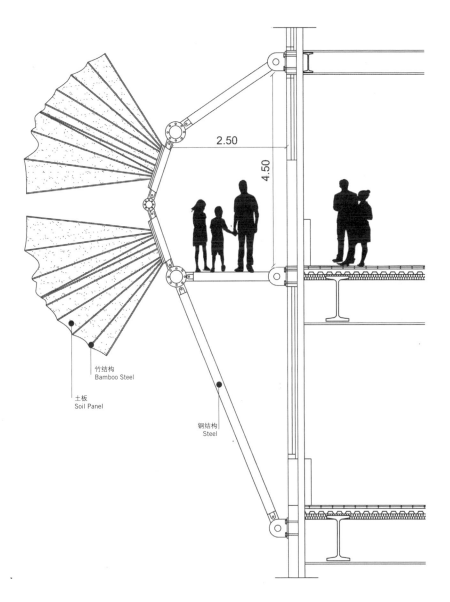

竹结构
Bamboo Steel

土板
Soil Panel

钢结构
Steel

装置移动方便，可伸缩，可开合。支撑结构类似于机器人手臂，可以弯曲伸展，吸附在建筑物表面。装置翅膀部分参照了蝴蝶翅膀结构，收缩与伸展模拟了蝴蝶破茧而出的过程。材料使用竹刚作为支撑，有高强度、低碳环保的优点；翅膀膜结构部分的材料是土板，上面种有吸附有害物质的植物。总体上可以将结构设计归纳为三点：亲近环境，减少碳排放；高强度，高密度，抗震结构；构件柔韧，结构系统完整。部分装置打开后，内部空间可以进入，其功能类似于高层的外挂阳台。

The device itself is portable and changeable and it could be folded and open. Supporting structure simulates robot arms, which could be bent and stretched and attach to the architecture façade. Ideas of the device's wings take structure of wings of butterfly for reference and refer to the process of a butterfly coming out of a cocoon. The material of supporting structure is wooden bamboo, which has the advantage of high strength and environmental friendly. Wings' material is soil panel, plants, with the ability to absorb hazily, grow on the panels. Features of structure design could be concluded into three points: Environmental friendly and low carbon consumption, High strength, intensity and anti-seismic; Pliable to machining and structure integrity. Some devices allow people to go inside like terraces.

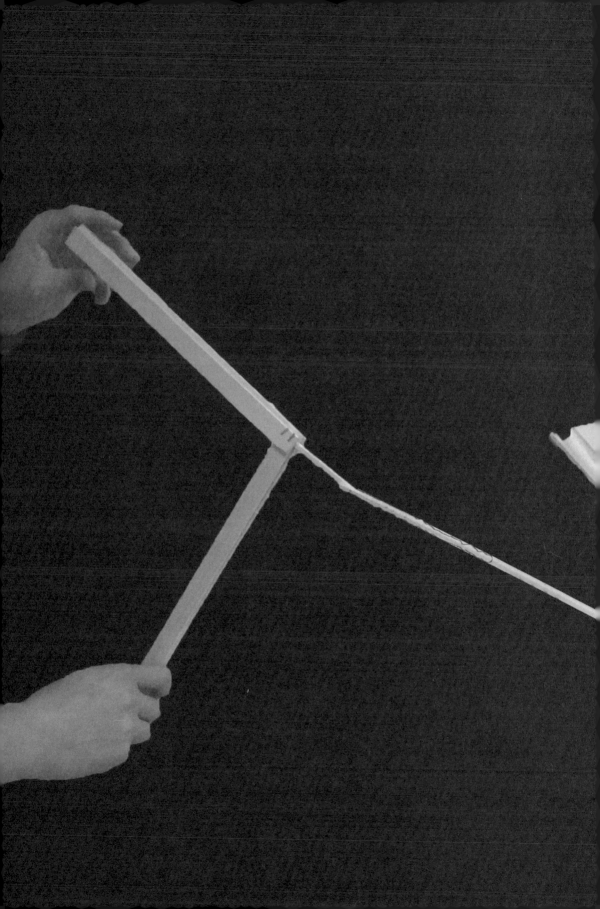

设计应对雾霾 DESIGN AGAINST SMOG

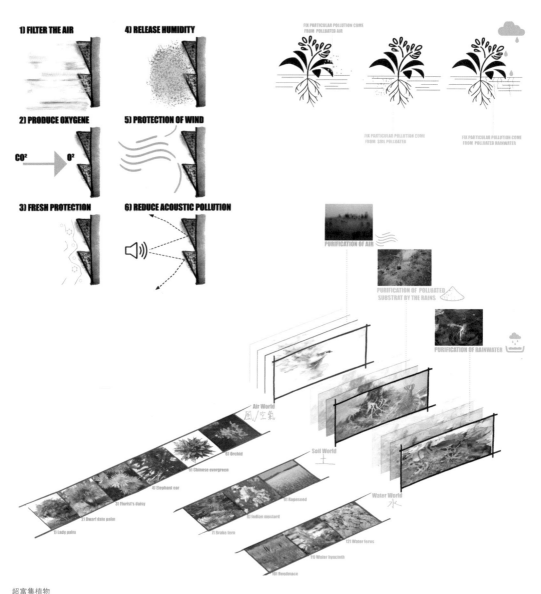

超富集植物
The Hyperaccumulators Plants

"一棵超富集植物能够在空气中、污染的水中、土壤中生长,通过它们的组织吸收这些污染物,并使之高度浓缩。"

"A hyperaccumulator is a plant capable of growing in air, or in polluted water, or in polluted soil with very high concentrations of pollution, absorbing this pollutant through their tissues, and concentrating extremely in high levels."

设计应对雾霾 DESIGN AGAINST SMOG

在上海高层最集中的区域，比如陆家嘴，这种区域高层林立，密度较大，且高层与高层之间存在涡流，容易聚集有害物质。"净化蝴蝶"装置集中依附在这些雾霾严重区域的高层幕墙表面以及高层阳台上。平时处于关闭状态，雾霾天翅膀会张开，使净化空气的植物土板暴露出来。

Lujiazui, gathering most high-rises in shanghai, is a very high density place. And some reversal flows (vortex) are observed here, in this circumstance most harmful substances are easy to produce and stay. "Purifly" attaches to the façade and terraces of high-rises in this area where big smog is serious, they stay closed at ordinary times and will be open in big smog days and expose the soil panels on which plants have the ability to purify air grow.

设计应对雾霾 DESIGN AGAINST SMOG

由于对城市环境问题严重性的忽视，城市的很多边界能源都没有得到利用。大都市中地铁和电梯的活塞效应，中庭和烟井的拔风效应能够带来不少被动通风，但这些都没有被利用。由于中国大都市的雾霾很大程度上都是由于自然气候条件中的静风环境造成的，在这个设计当中，我们希望利用城市的这些边界能源促进城市的通风。

The human activities' by-products, such as the piston effect of the metro and elevators and the stack effect of the dominating skyscrapers within the metropolis, are generally unexplored due to the ignorance of the severity of the environmental situation. Based on the fact that the smog problem within Chinese metropolis always concurs with the lack of air-flow due to meteorological reasons, this design intends to utilize these passive energies as the source of urban air-flow.

Core: Air Infrastructure
核：空气基础设施

Student 学生：
Tomoki Shoda 庄田智己
Liu Fangshuo 刘芳铄
Che Jin 车进
Wu Xiaoyu 武晓宇
Liang Qianhui 梁芊荟
Pablo Mariano Bernar Fernandez-Roca 巴勃罗·马里亚诺·博纳·费尔南德斯-洛加

Supervisors 指导：
Luo Jing 罗晶
Wang Zigeng 王子耕

核：空气基础设施 Core: Air Infrastructure

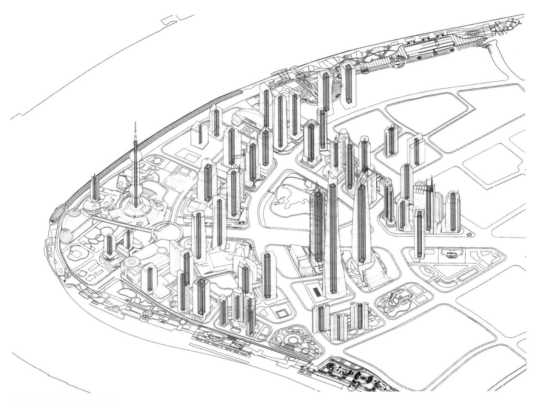

新的核心筒原型与陆家嘴更新
Cores in Lujiazui

设计应对雾霾 DESIGN AGAINST SMOG

陆家嘴：2条地铁线 / 700+ 部电梯
上海：16条地铁线 / 无数部电梯
每天有 400 万人使用地铁 (6 000 辆车)
每天有 1 100 万人使用地铁 (170 000 部电梯)

Lujiazui: 2 Subway lines / 700+ Elevators
Shanghai: 16 Subway lines / Countless Elevators
4+ Million People in Shanghai Travel by Shanghai Metro Daily (6,000 Trains)
11+ Million People in Shanghai Travel by Elevator Daily (170,000 Elevators)

核：空气基础设施 Core: Air Infrastructure

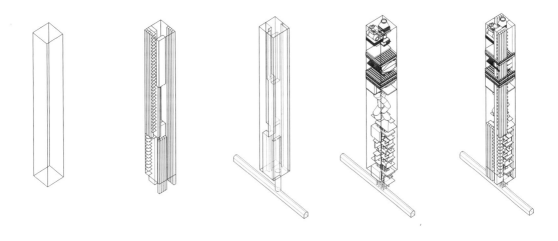

生成过程
Generation Process

通过对场地陆家嘴的认真分析，我们发现在变幻表皮的背后始终不变的高楼核心筒结构系统。倘若能够利用高楼的核心筒结构，将城市当中能够带来气流的这些被动能源进行整合，不仅能够帮助城市应对雾霾，更能给城市带来教育与启发的意义。

By a careful analysis of Lujiazui area, and the discovery of the never-changing core structure system behind the ever-changing facades of the skyscrapers, our team arrived at the conclusion that designing a new core prototype could be of great value not just to the incorporation of these passive energies mentioned above into the great war against smog, but also to the education of the entire population.

设计应对雾霾 DESIGN AGAINST SMOG

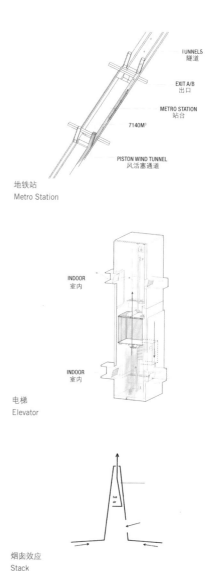

地铁站
Metro Station

电梯
Elevator

烟囱效应
Stack

除了充分尊重传统核心筒中必不可少的楼电梯、卫生间、管井等设施,我们希望在新的核心筒原型设计中融入当下城市最重要的元素——空气。来自于地铁、电梯、中庭、烟井的被动气流穿过一系列精心设计的空间和装置,空气当中危险的污染物通过离心、喷淋、环保膜、植被吸附、低压静电吸附等节约能源且成熟高效的方式处理掉了。

Besides the traditional functions of a core such as the stairs, the toilets, the shafts and the elevators, this new core prototype includes this very core urban issue of AIR. Passive airflows from subways, elevators, atriums and stacks are intentionally conducted through a serious of carefully designed spaces and devices, so that the dangerous pollutants in the atmosphere can be absorbed by the mature and energy efficient methods, including centrifuge, wet deposition, HEPA, phytoremediation, and low voltage adsorption.

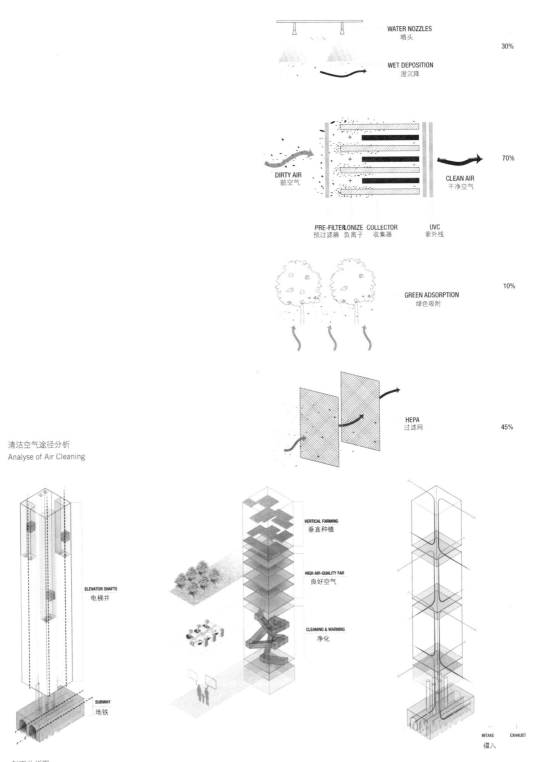

设计应对雾霾 DESIGN AGAINST SMOG

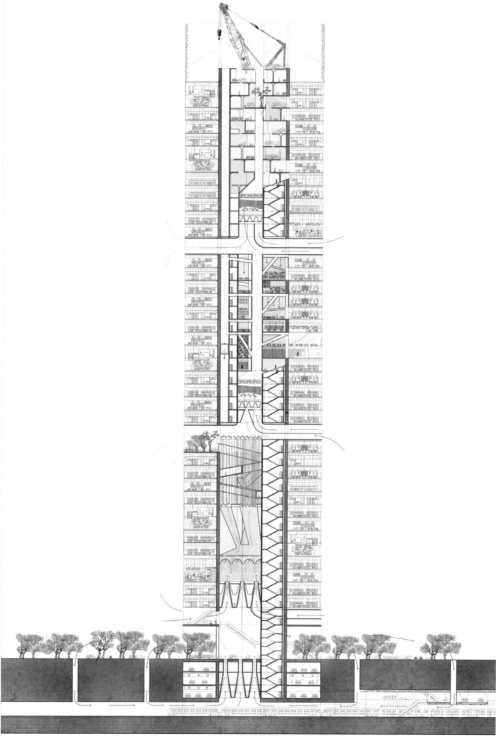

主剖面
Section of the Core

核：空气基础设施 Core: Air Infrastructure

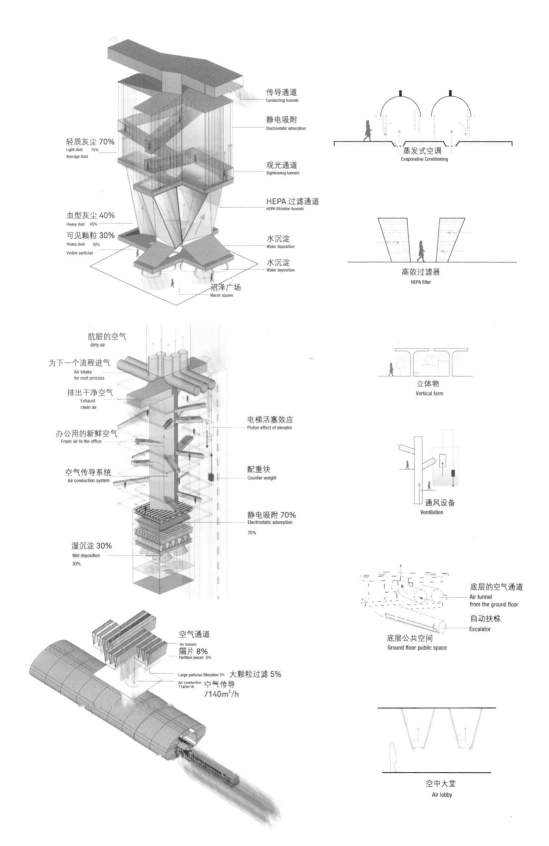

设计应对雾霾 DESIGN AGAINST SMOG

自上而下：教育 、空气博物馆 、垂直农场
Top to Bottom: Education, Air Museum, Farm

核：空气基础设施 Core: Air Infrastructure

应对雾霾问题需要政府与公民协同一致，远期而言，通过容积率政策等方式，或将能够促进开发商在新的建设当中更多采用新核心筒原型，新原型将城市利益考虑进来，使我们的建筑畅想更加可行。

一系列公共与绿色空间沿处理空气的全过程排布，使得建筑与建筑内外人群充分互动，增加公众对雾霾问题的关注意识。

Due to the smog becoming a national issue, the government and the citizens in China are forced to fight together. FAR policies can be adopted to encourage the developers to apply this new core prototype which benefits the city, making such bold architectural adventure more sensible.

More public and green spaces are created along this process, so that everyone within and without this building can interact with it to get more awareness of the air situation.

设计应对雾霾 DESIGN AGAINST SMOG

核心筒作为新的基础设施参与到城市更新的过程中
The new prototype of the core could participate in the urban regeneration process as an infrastructure

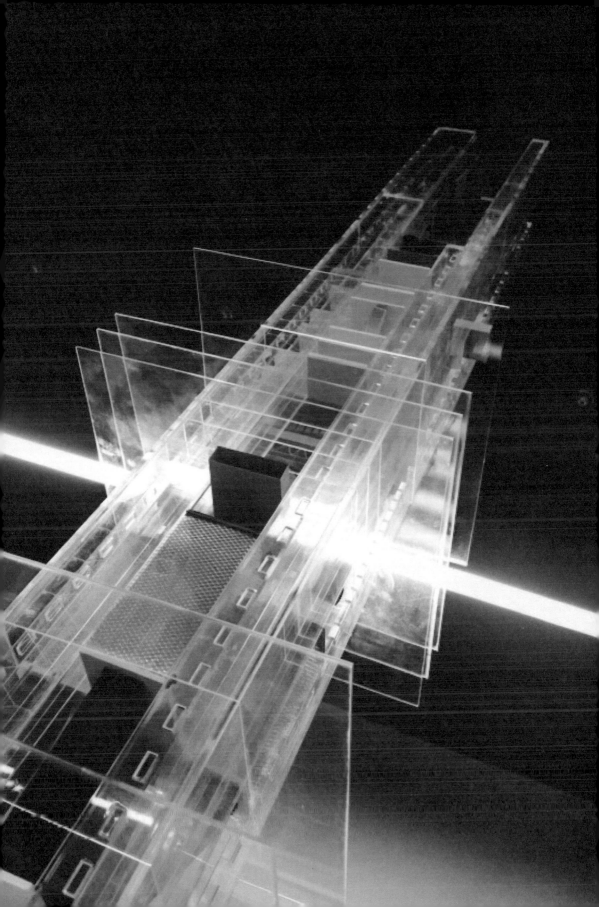

设计应对雾霾 DESIGN AGAINST SMOG

设计应对雾霾 DESIGN AGAINST SMOG

自空调发明以来，建筑就常常被设计为一个封闭的微气候系统，这一系统完全建立在化石燃料的消耗上。建筑师需要重新去发现传统城市灵活开放的环境策略，因为在那样的城市系统中，建筑的微气候与城市或区域的大气候是一个完整的整体，建筑通过主动的调控与外部气候系统之间的边界关系来适应并创造非对抗性的内部微气候。

从这一理念出发，我们运用热力学与复杂科学的思维方法，重构中国当代城市建筑的边界问题、物质问题、能量流问题与组织形式问题，并对现代主义的建筑体系进行深刻的反思。

Since the invention of air condition, buildings are often designed as a closed micro-climate system, which fully relies on the consumption of fossil fuel. Architects may need to return to the flexible open environment strategy adopted in traditional city, where the micro-climate of building and macro climate of city work in whole. in that case, buildings actively adjust the board relationship with external climate to adapt to the inner micro climate instead of against it.

Starting from this concept, we apply the thinking methods of complex science and thermodynamics to reconstruct the boundary problem, material problem, energy flow and organization form problems. Also, the modernism building system would be deeply reflected in this essay.

A Thermodynamically Driven Respiratory System
热动力呼吸系统

Student 学生:
Kang Da Hong 姜大宏
Li Ao 李骜
Jim Peraino 吉姆·柏雷诺
Yao Huiting 姚慧婷
Zhang Fan 张帆
Zhao Kong 赵孔

Supervisors 指导:
Lyla Wu Di 吴迪
Hu Chenchen 胡琛琛

热动力呼吸系统 A Thermodynamically Driven Respiratory System

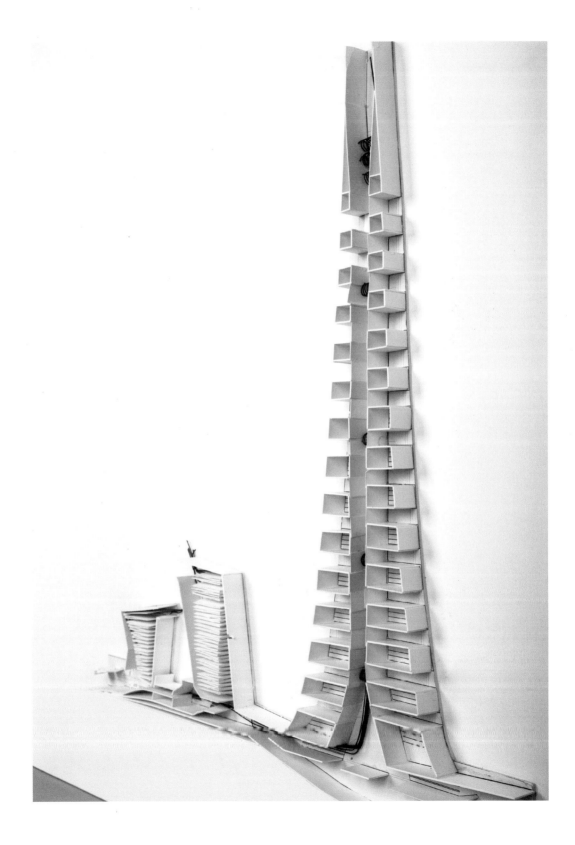

设计应对雾霾 DESIGN AGAINST SMOG

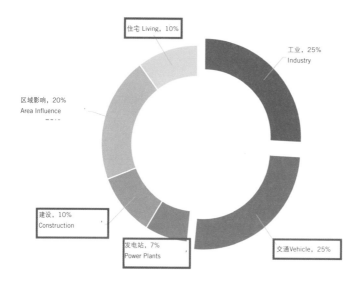

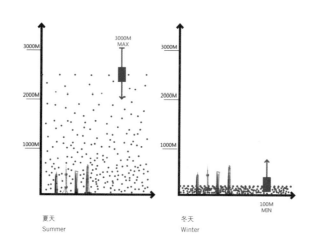

经过数据调查，我们小组发现陆家嘴地区有52%的雾霾与建筑环境有关。

夏天和冬天，空气的污染密度是不同的，我们暂且将这种区域面积定义为边界。在冬天，边界可以低至100米，导致空气中污染物密集且不健康。

The building environment can produce less smog. 52% of smog is related to the building environment.

During summer and winter, the air pollution density is different, We defined this area as the boundary. The boundary layer can get as low as 100m in winter, causing pollutants to be dense and making the air unhealthy.

热动力呼吸系统 A Thermodynamically Driven Respiratory System

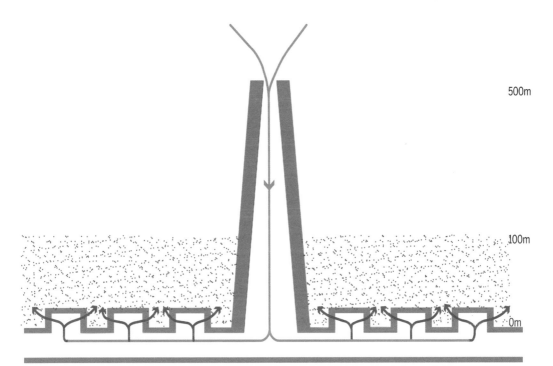

通过通高高塔，我们可以穿过边界将清洁的空气带到地表。
With tall towers, however, we can reach above the boundary layer and bring clean air down to the ground.

①也许每座塔像上海中心那么高，可以将清洁空气传送下来为自身服务。
Each tower could be as tall as the Shanghai Tower to bring clean air down for itself.

②或者利用现有建筑之间的空间将气流引下来输送到其他建筑。
Or we could use the space existing between buildings to support large downdraft chimneys that distribute clean air to other buildings.

③再或者，我们可以利用现有的超高建筑传送清洁空气下来为自己和其他建筑服务。
Or maybe we could clad existing tall buildings and use them to bring clean air down for all of the others.

设计应对雾霾 DESIGN AGAINST SMOG

BRING CLEAN AIR DOWN

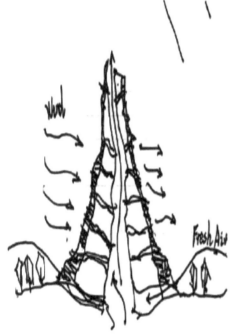

一旦冷空气到达设备的底部,建筑中的一个向心的热力核心筒启动,并且将空气运输向上。

通过理解了材料的性质(例如热扩散系数),我们可以确保热能的转换可以在我们需要的时间和地点发生。

Once cold air has reached the base of the chimney, a central "hot spine" of program jumpstarts the flow upwards.

By considering material properties such as thermal diffusivity, we can ensure heat transfer occurs only when and where we want it to.

热动力呼吸系统 A Thermodynamically Driven Respiratory System

街道网格如主动脉系统般封闭，将清洁空气重新分配到各个建筑中去。
The street grid enclosed and used as an artery system for airflow, redistributing clean air from the towers to each buildings.

这些塔使建筑之间形成网络，传输自然、洁净的空气。
The towers serve the community by bringing down fresh, clean air which can naturally ventilate the other buildings.

设计应对雾霾 DESIGN AGAINST SMOG

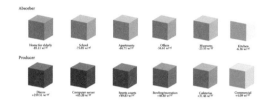

在夏天，所有的功能空间需要补充制冷、降温。
All programs require supplemental cooling in the summer.

即使在冬天，一些功能空间产生的热量大于消耗，因此还需要冷却。
Even during the winter, some types of programs produce more heat than they use, necessitating cooling.

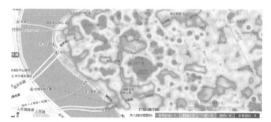

此区域缺乏功能的多样性，所有的功能活动都集中在白天。
Currently, there is a lack of diversity uses. All of the functional activities are concentrated in the daytime.

夜间却几乎没有可以使用的功能。
At night there almost no function can be used.

热动力呼吸系统 A Thermodynamically Driven Respiratory System

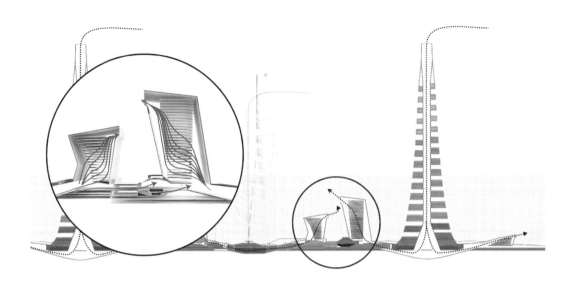

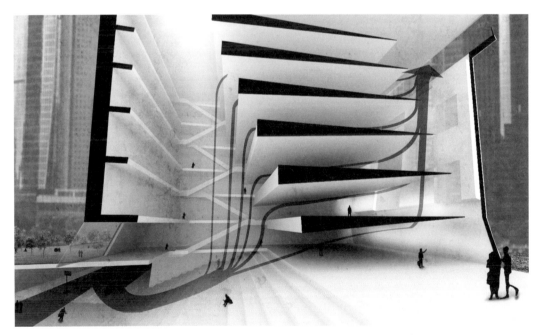

在剖面上，我们适当地调整了楼板的形态，可以更好地让气流形成循环。

On profile, we propose adjusted the shape of the floor plate ,to create air circulation.

设计应对雾霾 DESIGN AGAINST SMOG

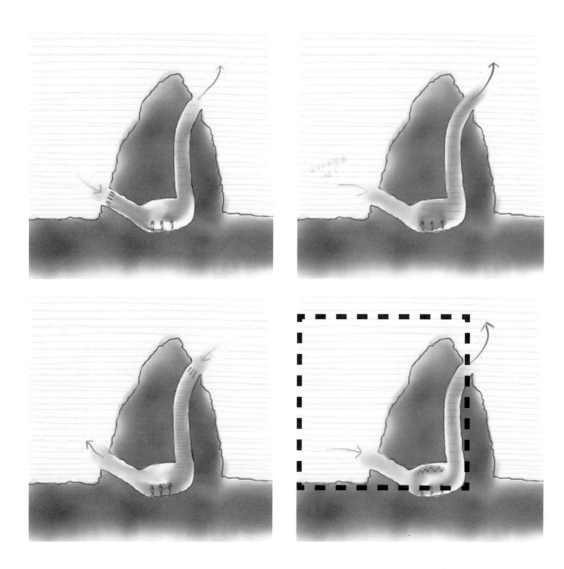

我们可以利用温差结合建筑的组织形式来控制气流和空气的走向。室内的活动导致空气加热,当这股气流上升时,它排出建筑物,导致新的气流从上方进入建筑内部取而代之。

Instead, we can use temperature differentials coupled with building organization and form to control the flow of air. Interior activities cause the air to heat up. As this air rises, it flows out of the building, causing new air to come in from above to replace it.

热动力呼吸系统 A Thermodynamically Driven Respiratory System

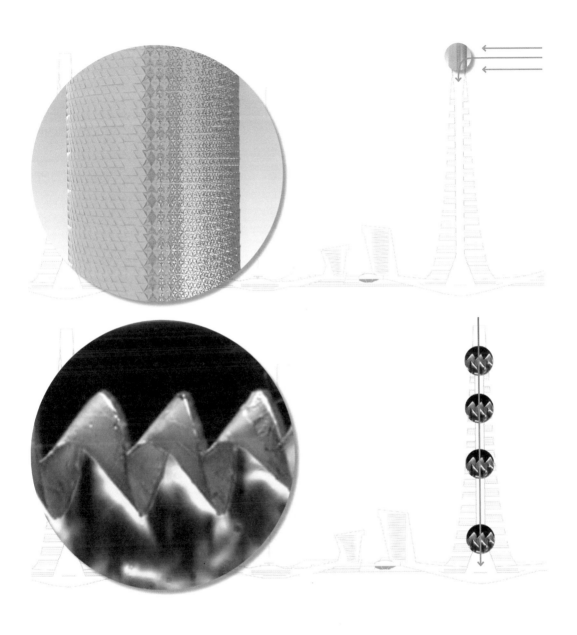

在塔的顶端采用可调控开合的表皮，这样确保在不同的季节，下论什么方向的气流都能被引导下来。

一些机械冷却设备是必要的，当气流进入设备的热活性表面时，由此可以控制气流。和传统空调相比，这些设备将效率提高了。在塔里的不同高度设置装置以更流畅的疏导干净空气向下流动。

A responsive skin at the top can adapt to wind direction, ensuring that air in the chimney always flows down.

Some mechanical chilling equipment is necessary. When air flow to the thermo-sensitive surface of the equipment, they can control the flow. Compared to the traditional air conditioner, this equipment will improve the efficiency. These equipments can be installed in different height inside the tower, in order to transfer clean air downwards.

设计应对雾霾 DESIGN AGAINST SMOG

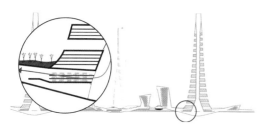

在冬天,气流先通过底层的核心的电脑服务区,这里产生大量热能,可以在到达居住区之前温暖气流。
During the winter, air can be directed through computer servers at the base of each chimney which produce excess heat, warming the air before reaching inhabited space.

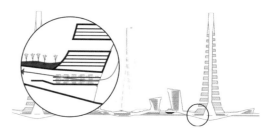

天花板有一定的倾斜,这样当热流通过核心筒分流时可以被分流。
Ceilings are sloped so that it is directed away from the chimney, as the hot air rises.

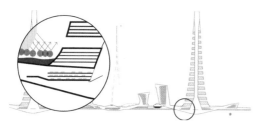

在夏天,气流可以被再导入以保持凉爽。
During the summer, air can be redirected to stay cool.

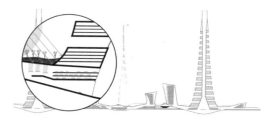

热气团在冬天时,于核心筒底部下方进行加热。
Thermal mass at the base of the chimney can heat the air below it during the winter.

在夏天,树可以阴凉热气团,这样有助于冷却空气。
Trees can shade the thermal mass during the summer, so that it helps cool the air.

最后,整个陆家嘴区域形成一个整体的、开放性的、建筑间彼此配合的组合关系。使人们最大可能的不受雾霾的影响,让生活于钢筋水泥丛林的现代人得以放松身心。营造充满亲切感的里弄街巷,拿士拘束的室外空间,亦创造出一个多元交流的场所,人与自然、人与建筑、人与历史文化、人与艺术,以及人与人的对话将更加直接、有趣。

Finally, the Lujiazui area is to form a integral and open space, with buildings cooperating with each other. Most people live in concrete jungle can avoid the harm from hazardous fog and easy their minds. The streets and alleys with intimacy, the outdoor spaces with no restrain, the communication channel for people from diversified cultures are then created. The communication of people and nature, people and buildings, people and history, people and art even people with each other, would be much more direct and interesting.

设计应对雾霾 DESIGN AGAINST SMOG

中国的雾霾因为工业野心和几十年的车辆排放变得更为严重。由于缺乏政策的落实和执行,设计人员面临着设计一个更加光明的未来的任务。

经过对当前状况的广泛分析,我们的团队利用不断增长的网络和实质性的响应系统,从源头解决烟雾问题。"藤"主要针对空气污染在该地区的两大主导来源:汽车和工厂排放。这种温柔的干预将减少小汽车的使用,创建一个行人和自行车友好的系统,从而使生活方式发生改变,使用模块化的预制基础设施,极为方便。它利用能够分解的二氧化钛,可以使用混凝土或涂料作为主要处理有害颗粒物的材料。这样的设计也旨在最大限度地提高热舒适性和整个陆家嘴地区的透气性,提供给当地人和游客一个愉快的环境。

如上海继续在未来发展,该网络可以在整个城市延伸,以适应不同的需要和条件。藤提供了一个长期的,全面的解决方案,能够与城市一起成长。通过实施这个系统,上海的未来可以出现更健康的公共空间,更好的连接性,成为一个蓝天白云并存的上海。

China has been suffering from smog as a result of industrial ambitions and vehicle emissions over several decades.

After extensive analysis of current conditions, our team has taken an approach to use a responsive system of growing networks and materiality to combat smog. The Vine targets the leading sources of air pollution in this district: car and factory emissions. This gentle intervention will support a change in lifestyle in order to reduce car use by creating a pedestrian and bike friendly system. The modular infrastructure can be prefabricated for easy and adaptable construction. It utilizes TiO_2, a material that breaks down harmful particles that can use concrete or paint as a host for flexible application. This design also aims to maximize thermal comfort and ventilation of the Lujiazui District to provide locals and tourists an enjoyable environment.

As Shanghai continues to develop in the future, this network can extend throughout the city to adapt to various needs and conditions. The Vine provides a long term, holistic solution that is able to grow with its city. By implementing this system, the future of Shanghai can look forward to healthy public spaces, greater connectivity, and a blue, smog-free sky.

THE VINE
藤

Student 学生:
Davide Masserini 大卫·马萨里尼
Veronica Gazzola 维罗妮卡·佳佐拉
Alyssa Maristela 艾丽莎·玛丽斯黛拉
Zhang Zhenwei 张振伟
Wen Zishen 温子申
Sun Tongyue 孙童悦

Supervisors 指导:
Minqu Michael Deng 邓闵衢
Yang Feng 杨峰

设计应对雾霾 DESIGN AGAINST SMOG

曼哈顿
Manhattan

米兰
Milan

伦敦
London

巴黎
Paris

密度比较研究
Density Comparison Studies

模型照片
Model Photo

城市系统
Urban Systems

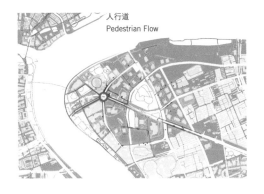

徒步慢走系统和自行车系统在陆家嘴地区非常需要。

An improvement of the pedestrian and bicycling system is greatly needed in the Lujiazui District.

设计应对雾霾 DESIGN AGAINST SMOG

日光分析
Sunlight Analysis

风分析
Wind Analysis

雾霾分析
Smog Analysis

空气污染的情况与太阳和风的存在形式有关，考虑到改变这些环境元素对于提高陆家嘴的空气质量非常重要。

The condition of the air pollution is related to the presence of the sun and wind. Studying these environmental systems becomes important in creating a design that considers how these elements can improve the air quality in Lujiazui.

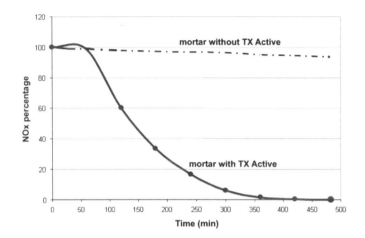

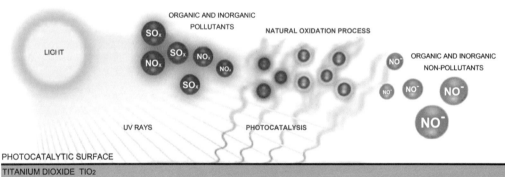

吸收雾霾材料
Eating Smog Material

研究一种新型的建筑材料产生二氧化钛，使用新型手段和元素产生新的可以用于光合作用的材料来破坏有害颗粒物。

Now research of building materials that combat pollution has resulted in the application of titanium dioxide (TiO2), an element that uses a process similar to photosynthesis to break down harmful particles.

设计应对雾霾 DESIGN AGAINST SMOG

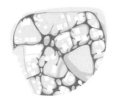

底层自行车路径
Lower Level Bike Lanes

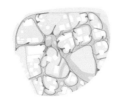

底层步行道路径
Lower Level Pedestrian Spaces

上层自行车路径
Upper Level Bike Lanes

上层步行道路径
Upper Level Pedestrian Spaces

底层雾霾
Lower Level Smog

上层雾霾
Upper Level Smog

日光
Sunlight Levels

风
Wind Levels

藤是一种双重高度的行人天桥系统，这将提高自行车和行人的体验。通过改善整个陆家嘴地区的连通性，当地人和游客都被鼓励使用步行、自行车和公共交通。藤的两层设计是为了将用户和车辆分开，给他们提供安全和移动的方便。

在两层之上的穿孔式伞盖式层为人们提供遮阴，自然通风同时能除去空气中有害物质的材料。这种干预能够很好的延伸至整个城市来适应上海的发展。

The Vine is a double height, elevated walkway system that will enhance the bicycle and pedestrian experience. By improving the connectivity throughout the Lujiazui District, both locals and tourists are encouraged to commute by foot, bike, and public transportation. The two proposed layers of The Vine are designed to separate users from the vehicular level, increasing their safety and ease of movement.

A perforated canopy layer above both levels provide shade for people, the passing of natural ventilation, and the material that breaks down the air's pollutants. This intervention can be easily extended throughout the city to adapt to Shanghai's growth.

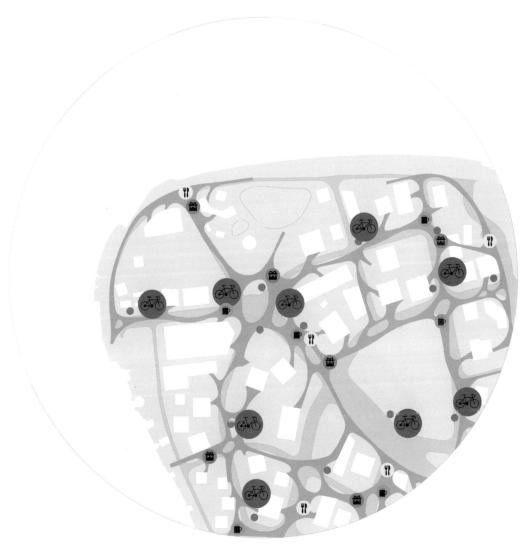

功能分析
Programs

为了进一步完善藤的设计，我们已经将激活功能如咖啡，食品小亭，和小商店等沿层设置。

在街道上，自行车中心被放置在一个舒适的步行距离可到达的现有公交车站内。这些中心提供自行车租赁服务以及个人自行车停放空间，形成了一个简便的公共交通方式转换的地点。

In order to further enhance The Vine, we have allowed space for activators such as small kiosks of coffee, food, and small shop vendors along the layers.
On the street level, bicycle hubs are placed within a comfortable walking distance of the existing bus stops. These hubs provide bike rentals as well as space for personal bikes, making an easy transition for users of public transportation.

设计应对雾霾 DESIGN AGAINST SMOG

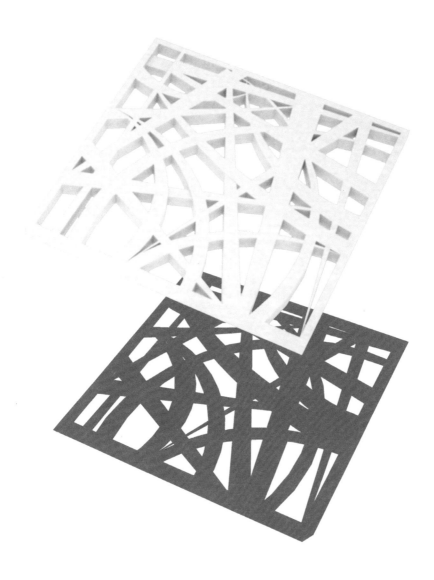

材料的应用
Material Application

一种木质复合结构,外表将涂有含有二氧化钛的涂料。穿孔设计增加了表面积与体积之比,从而增加材料的光催化作用效率。它也使自然通风能贯穿整个系统。

Made of a wood composite structure, the skin will be finished with a paint containing TiO_2. A perforated design increases the surface area to volume ratio, thereby increasing the efficiency of this photocatalytic material. It also allows natural ventilation to pass throughout the entire system.

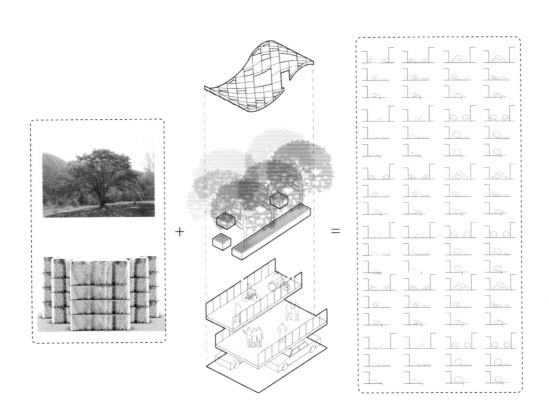

单元组合
Element Combination

设计应对雾霾 DESIGN AGAINST SMOG

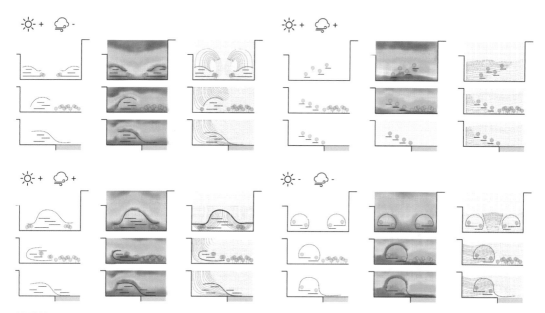

剖面分析
Section Analysis

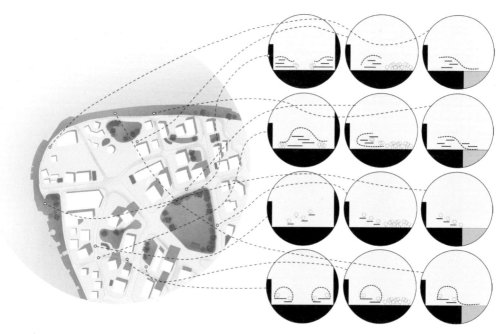

剖面组合
Section Composition

藤 The Vine

模数研究
Modular Research

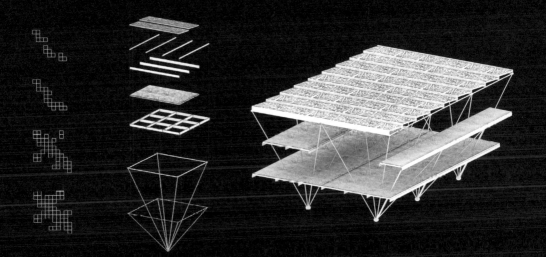

结构轴测
Structure Diagram

设计应对雾霾 DESIGN AGAINST SMOG

典型剖面
Typical Section

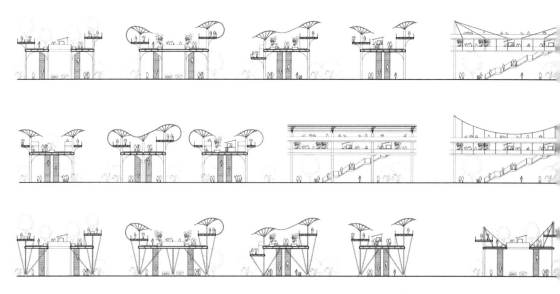

剖面类型
Section Typologies

190-191　　藤 The Vine

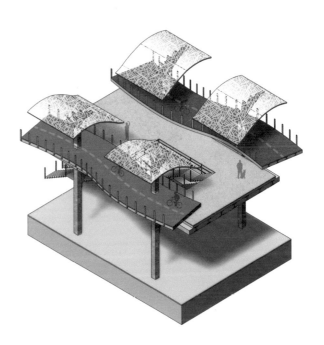

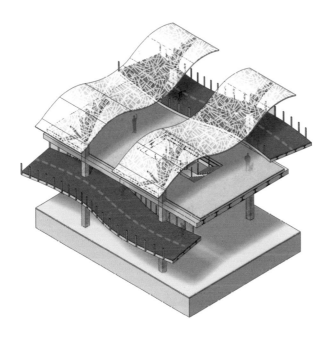

等角剖面
Isometric Section

设计应对雾霾 DESIGN AGAINST SMOG

总平面
Master Plan

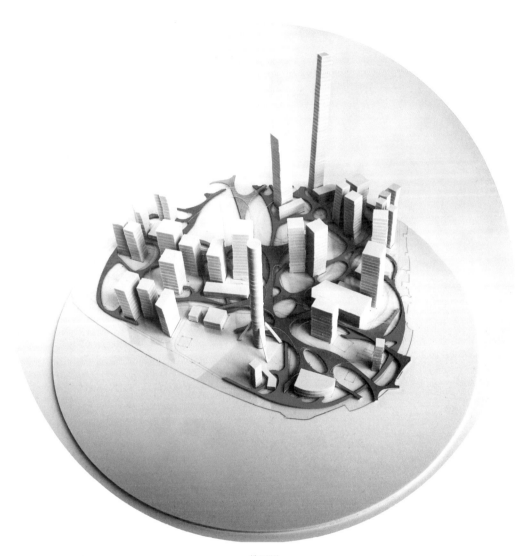

模型照片
Model Photo

设计应对雾霾 DESIGN AGAINST SMOG

城市的发展日新月异,在全球化流动空间的背景下,信息技术给一切带来了自下而上变化的可能,空间生产的模式也不再一如既往。我们提出一个ESC平台来引领这种自下而上的绿色空间的全民生产,以陆家嘴区域为实践的首发之地,众智、众筹、众包、众创治理雾霾——大大小小的各种尺度的武器将会被投放到场地对雾霾发起抵抗,这是一场全民运动,是一场牵涉设计师、厂商、投资方、群众等的全民运动。新生产的绿色空间根据热力学原理在平台上进行集思广益的规划布置,不仅能促进节能减排,还能获得经济效应与社会价值。

ESC (Energy System City) is a platform proposed for the Lujiazui District where everyone can come together as a whole to help solve the issue of smog. This platform creates a place for any solution to be proposed for the district, both big and small scales projects, and the people decide on the best proposal through crowd funding, which is integrated with the platform. ESC proposes different solution to both address the issues of energy production and reduction. ESC is a place for designers, developers, government officials, and society to contribute to creating a better district. This platform starts in Lujiazui but can be used throughout all the world for everyone to address this issue.

STEP:ESC SMOG
步骤:退出雾霾

Student 学生:
Amalia Checa Gimeno 艾米丽娅·切卡·季米诺
Evan A Weaver 伊万·A·韦弗
Ling Mengzhi 凌梦芷
Zhou Yifan 周一凡
Zhang Chengyuan 张程远
Yan Yu 言语

Supervisors 指导:
Su Yunsheng 苏运升
Georgiev Georgi 乔治·吉奥吉夫

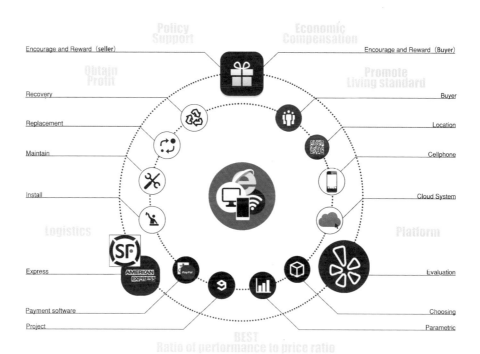

PEST 的组成元素
Elements of PEST

STEP 的作用机制
Mechanism of Work of STEP

设计应对雾霾 DESIGN AGAINST SMOG

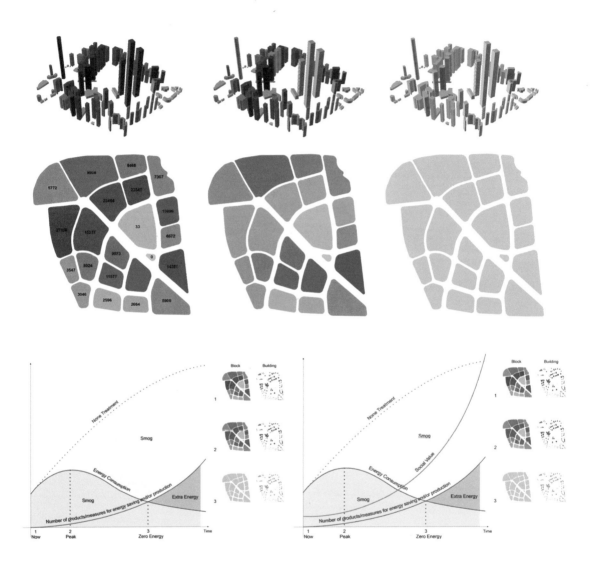

我们运用软件来分析了陆家嘴片区的建筑地块能耗。颜色越深代表能耗越高。我们模拟了能耗逐步降低并趋于平衡的三个阶段，并通过图表来分析绘画出采取措施前后的能耗曲线，雾霾曲线，社会价值曲线。

We use software to analyze the energy consumption of Lujiazui building distractions. The deeper the color is, the higher the energy consumption is. We try to simulate three stages of the decreasing and the balance of energy consumption and draw the chart to illustrate if we take some measures to Lujiazui area, what will change in social value, the energy consumption, and smog.

cold current
Heavy rains GAME fierce wind
flood
intense heat of summ
CROWDFUNDING El Nino
typhoon SMOG
PROJECT

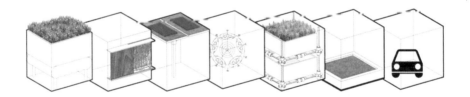

相对于严肃的基金整合平台，类似于游戏性质的众筹平台载体，具有亲和性高、用户黏性高等特点；针对目前公众难以参与城市规划方案的痛点，通过快速反馈有效增强玩家同众筹平台的联系。由游戏平台上的众筹项目，发展到真实可行的现实众筹项目，从而促进政府、设计师、公众、公司等进行多方合作。

Compared with serious fund integration platform, the crowd funding system with game basis has higher affinity, higher user stickiness characteristics, which can effectively focus on urban planning pain point where that citizens normally cannot participate in. It can also receive the feedback from all the game players through the connection system. Furthermore, from the game crowd funding system to raise the real project in city and at the same time it can promote cooperation between goverment, designers, public, and companys.

设计应对雾霾 DESIGN AGAINST SMOG

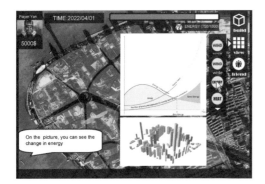
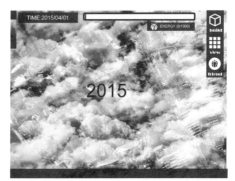
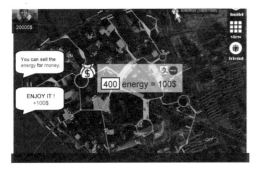

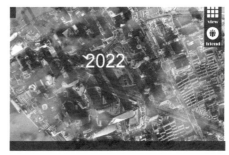

以游戏的方式诠释
Illustrated as a Game

游戏众筹平台中的所有工具包均来自设计师之手,在一定程度上能够减少雾霾,美化环境。所有的工具包的应用均由政府批准,设计师设计,厂家生产,制造商安装制造,民众评价并且反馈意见到众筹平台上,政府再通过这些反馈判断这些工具包的好坏。

All the tool boxes in this crowd funding game system are designed by designers can deduce the smog to some extent and also have a good effect on beautify the city in the angles of urban planning. All the tool boxes need government's permission, designer's idea, producer's production and also need manutacturers. Furthermore, people can put forward their needs and evaluation on this platform, and government can decide which tool box is good or not good enough based on people's foodback.

设计应对雾霾 DESIGN AGAINST SMOG

城市系统的能量图释
Energy Diagrame of Urban System

设计应对雾霾 DESIGN AGAINST SMOG

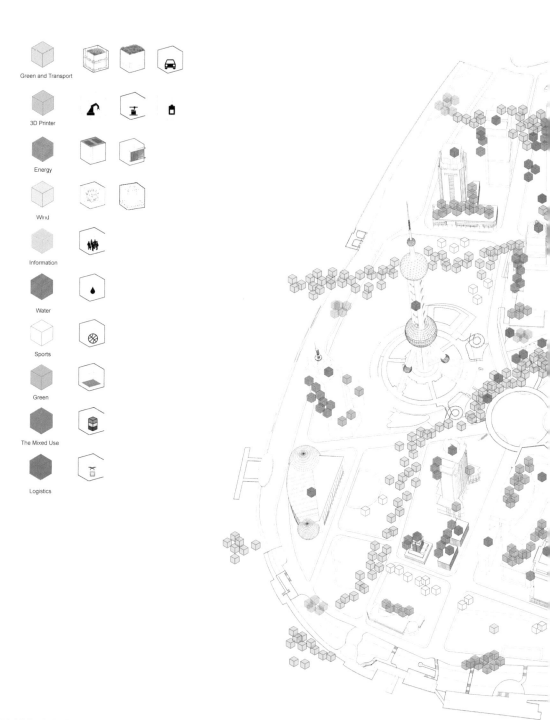

城市系统中的能量部件分布
Distrbution of Energy Components in Urban Systems

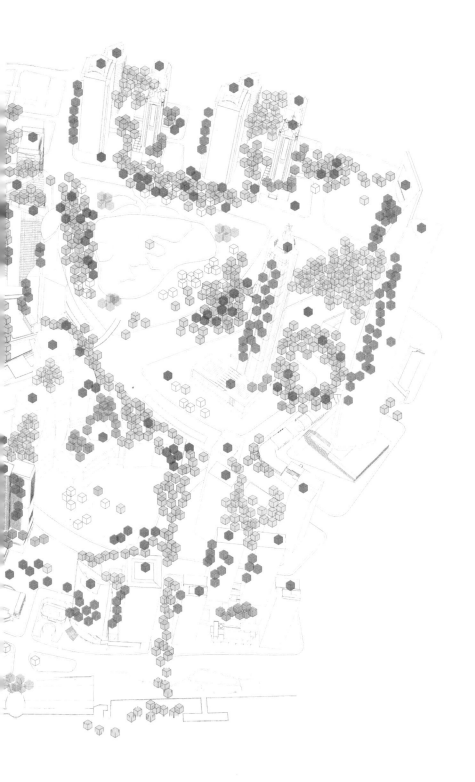

设计应对雾霾 DESIGN AGAINST SMOG

众筹桥梁
Crowdfunding Bridges

图书馆
Liberay

气球
Balloons

步骤：退出雾霾 STEP:Esc Smog

众筹花园
Crowdfunding Garden

漂浮的绿地
Floating Green

设计应对雾霾　DESIGN AGAINST SMOG

与其将雾霾问题当做一个特定基地或项目的独有问题，不如通过重新定义一个平行且相互影响的环境互换系统来应对雾霾问题。这种交互不仅可以被理解成为症状与原因之间交互，也可以理解成为雾霾的生产者和消费者之间的交互。因此本次设计将同时考虑产生雾霾的工厂以及建设能够消耗雾霾的基础设施。

我们的目标是控制并高密化城市中心的发展，整合城市基础设施，使其同已建成的环境协同发展，从而充分利用现存的资源。然而扪心自问，在这样的城市里，可以具体在哪里展开呢？由于我们感兴趣的是建立一个不仅适用于上海，同时也适用于中国其他城市的模型，我们关注这些城市之间的共性。最终我们将兴趣点落在它们与河流之间的存在关系。

Rather than identifying the smog issue as a stand-alone problem of a given site or program, we tackle the condition of smog through reconceptualizing a parallel and mutually reciprocal system of environmental exchange. This exchange can be understood not only in terms of symptoms vs. causes, but also consumers vs. producers of smog. As such the project proposes to look at both the industries that produce smog as well as the built infrastructure that consumes it.

Our goal is to control and densify the growth of the city within the city center, integrateing the infrastructural systems of the city to work alongside with building environment to take advantage of the inherent energies of each one. As we asked ourselves, in a city like this, where can we grow into? And since we were interested in creating a model that would not only apply to Shanghai but to other growing cities within China we looked at some of these and their commonalities, and we were particularly interested in the presence and relationship to the river in many of them.

RECY-Procal Urbanisms
互惠的城市主义

Student 学生：
Sofia Blanco Santos　索菲亚·布兰科·桑托斯
Fuyu Miyamoto　宫本冬
Kang Shinwoong　姜信雄
Pi Xin　皮歆
Huang Di　黄迪
Mao Yujun　毛宇俊

Supervisors 指导：
Liu Yuyang　刘宇扬
Oscar Ko　高亦陶

设计应对雾霾 DESIGN AGAINST SMOG

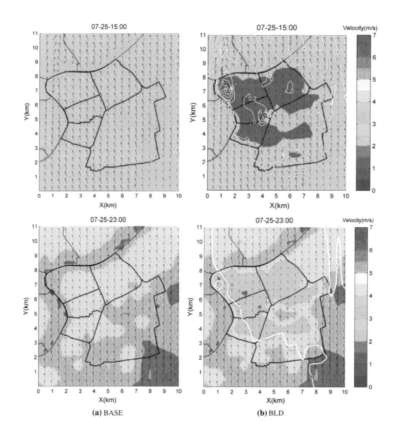

(a) BASE (b) BLD

从热力学和能量方面来说，这些城市中的河流都蕴含着极少被利用的巨大能量。在我们看来，这里正是城市中风能最为集中的地方，同时也有水流本身的能量。

In thermodynamic and energetic terms, these rivers in the cities are lines with an enormous degree of energy that in many cases is not taken advantage of. As we can see here is the part of the city where most energy of the wind is concentrated, as well as energy from the movement of the water itself.

设计应对雾霾 DESIGN AGAINST SMOG

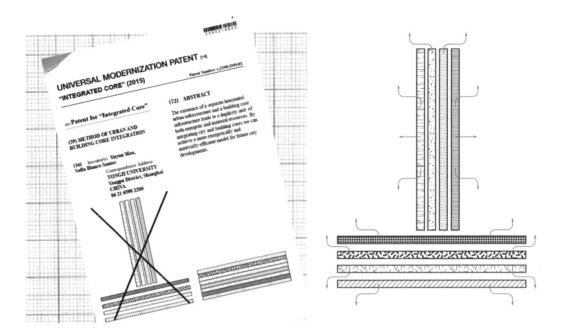

这促使我们开始想象城市在黄浦江江面上的生长方式。与在陆地上建造高层一样，我们设想将这些塔楼和他们的核心筒放倒，水平放置于江面上。这样做会抛出一个问题：我们该如何避免建筑基础设施核心和城市核心区域的双重性？因而我们开发了一个名为"集成核心"的原型。

This lead us to imagine how the city could begin to grow within the Huangpu river. But rather that building towers up as if we were on land, we thought of the possible consequences of laying these towers and their cores horizontally over the river. This would in turn pose the question of how we can avoid the duality between the infrastructure of the building core and that of the city core, So we developed a prototype for what we called the Integrated Core.

waste points

leisure and urban program infrastructure *recycling centers*

waste cycle

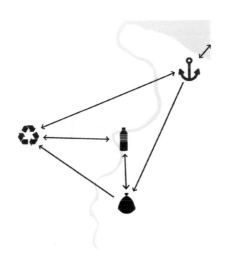

早先的垃圾处理流程依赖于交通运输,是由于垃圾站、冶炼工厂、回收中心以及各种产业分散在不同区域。因此,该方案设想了一个回收中心,可以将废物处理的所有程序在此集中,以减少交通运输量。

The original waste disposal flow relies on the transportation, since the garbage stations, smelting factories, recycling centers and industries are scattered in different areas. Therefore, the scenario here imagining a recycle hub that combines all the programs needed to waste disposal in order to reduce transportation.

设计应对雾霾 DESIGN AGAINST SMOG

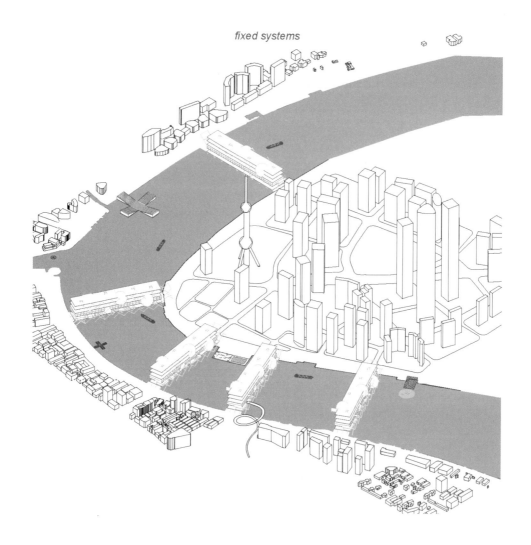

固定系统
Fixed systems

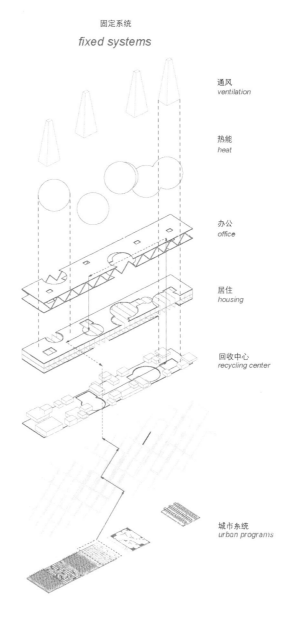

"集成核心"建筑用有不同的功能组织程序。从底层到高层,例如:传输,水力能源生产运输,连通性废物处理,回收利用中心,住宅与公共项目等。

The Integrated Core building has several layers of different programs. From lower-level to the top-level, that is : transfer, hydraulic energy production transport , connectivity waste treatment, recycling center, housing and public programs.

设计应对雾霾 DESIGN AGAINST SMOG

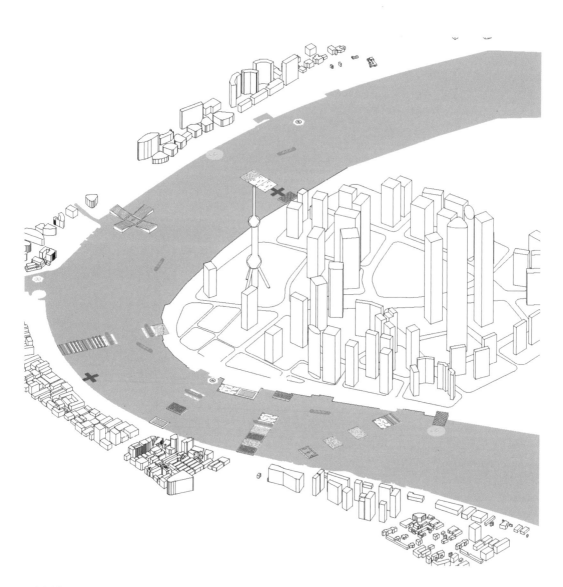

可移动系统
Movable systems

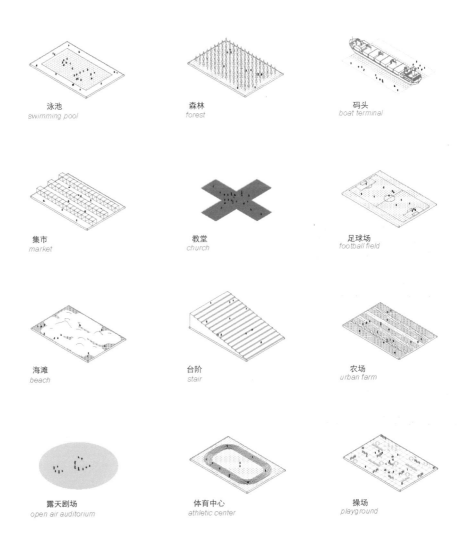

最终，桥内设置一些会产生大量垃圾的功能，例如：
_ 市场
_ 公园
_ 体育馆
_ 音乐厅
_ 移动游船码头
_ 游乐园

最终，这些桥的存在和这些新的项目间吸引力，将会吸引人群流入陆家嘴地区，引领更大的发展前景，带来更高的密度以及区域功能的混合。

Finally, programs that generate larger amounts of waste are placed within the bridges. These are:
-Market
-Park
-Sports Halls
-Concert Spaces
-Cruise Terminal for Boats
-Fun Park

As a result of the existence of the bridges and these new programmatic attractions, there will be an influx into the Luijiazui district, leading to greater development, density and mix of uses in the area.

设计应对雾霾 DESIGN AGAINST SMOG

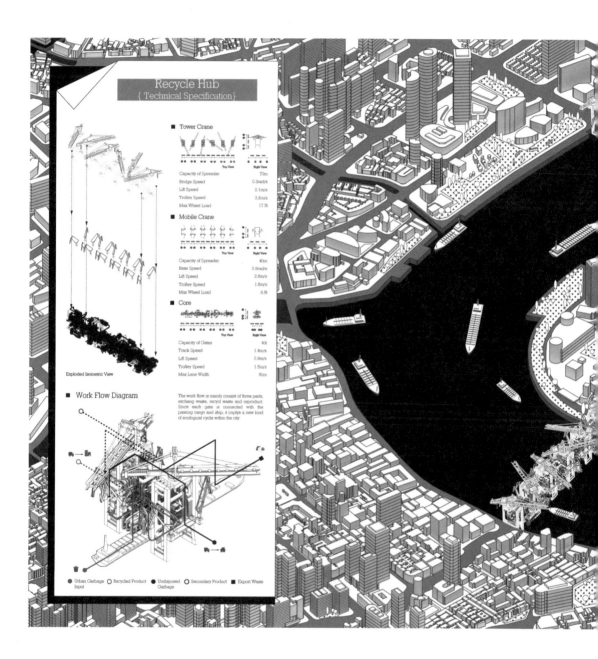

互惠的城市主义 RECY-Procal Urbanisms

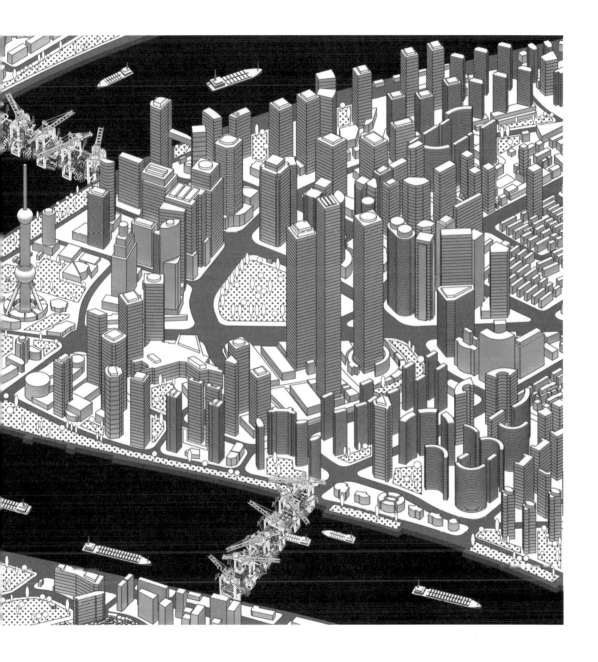

附录
暑期学校纲要

20世纪，人类创造出空前繁荣的物质文明，同时严重破坏了地球生态环境和自然资源。城市、建筑与环境之间的矛盾已经日益严峻，同时也给城市的发展带来了巨大的压力和阻碍。在建筑学日益关注社会议题的今天，传统建筑学的自主性受到了巨大的挑战，建筑师们陷入了生态技术和理论上的焦虑。在这样的困境下，知识架构转型就成了必要。

2015年同济CAUP的暑期设计学校活动主题为"设计应对雾霾：热力学方法论在中国"，旨在探讨及尝试用热力学的方法、思路和范式来响应当代中国的环境危机，力图在更大的议题中寻找建筑的自主性。

1. 雾霾

2015年初，自由新闻人柴静的纪录片《苍穹之下》，消解了公众了解雾霾问题的技术壁垒，也促使规划、建筑、制造与能源等领域人士去构建一个整合多种专业的方法论框架，进而对当前中国城市与环境问题进行分析、诊断。值得注意的是，第二次世界大战后的西方世界亦经历了类似的能源与环境危机：20世纪50年代伦敦雾霾污染事件，20世纪60年代洛杉矶光化学烟雾问题，至今仍饱受雾霾困扰的巴黎——试图通过汽车限号的方法来减少雾霾天气。

2. 空气

何为空气？是一团巨大的、虚无的空洞，还是这个世界的透明溶剂？

我们看不见空气，却可以感受它的力量。空气能拂过草地掀起涟漪，能吹动浮云，能在海岸边卷起浪花。

空气微粒也有层次，它的透明性掩盖了它的复杂与丰富。空气颗粒甚至可以被操纵成不同的纹理、表面和空间。

我们以为自己呼吸的是"当下的"空气，但实际上空气是流动的，它具有非均质性与非同时性——我们吸入的其实是源自不同地点，甚至是不同时期的空气。2008年的"In the Air"是一个关于空气可视化的项目，通过可视化工具让马德里的微观空气（尾气、颗粒物、花粉、病毒等）可见，使得人们能观察到空气颗粒是如何作用、反应，如何与城市界面进行交互影响的。

因此，在呼吁改善气候与环境的大背景下，建筑师们也做了许多尝试，不再只注重建筑的形象与功能，而是关注气候与转译，致力于空间、空气及其流动，以及传导、出汗、凝结、波动等气象状况，而这些将会成为当代建筑的新话语。"何为大气，是空气还是气候？或者两者之间——效应、物质、非物质、空间和现象"？

早在1959年，巴克敏斯特·富勒在纽约曼哈顿岛上空的大穹窿构想，就映射出高速现代化时期的西方对环境的急剧关切。在这样一个时代，建筑将不得不面对环境危机的挑战，不得不向富勒的技术文化遗产学习。

3. 热力学

自1865年鲁道夫·克劳修斯提出热力学第二定律以来，"熵"作为一种重要的系统状态参数被引入物质的组织形式研究，100年后，伊利亚·普里高津为关于耗散系统以及不平衡系统的自组织研究奠定了坚实的基础。据此观点建筑可视为一个非平衡态的不可逆过程的热力学"耗散结构"。这一耗散体以最大化的能量交换和熵的维持为特征，必须在一个热力学整体系统中加以考量。

热力学与复杂科学将城市与建筑看做一个边界开放的物质与能量的组织系统，这对现代主义决定论的、封闭的、技术至上的城市观是一种革命。空调的发明使建筑时常被设计成为一个完全建立在化石燃料消耗基础上的封闭微气候系统。如今，建筑师要重新发掘传统城市灵活开放的环境策略——建筑的微气候与城市或区域的大气候作为一个整体考虑，建筑通过主动调控与外部气候系统之间的关系来适应并创造非对抗性的内部微气候。

热力学建筑试图基于能量流动与形式生成的研究重建一种建筑批评与实践范式。这不仅仅关乎建筑学本身，更是基于百年来建筑与科学的发展图景，通过跨学科知识的考古、话语重构和工具开发，融合了本体性和工具性，提供新的建筑类型、形式与范式。彼得·卒姆托在奥地利布雷根茨的博物馆，通过将墙板中的水冷系统与通风系统分离，清晰地展现了建筑的建构特征；菲利普·拉姆则通过一系列图解来试图呈现热力学"百科全书"，他的艺术家住宅生动地展示出消解的空间边界；伊纳克·阿巴罗斯、基尔·莫·桑福·昆特等人在哈佛大学设计研究生院展开的"热力学建筑"实验，是通过知识体系、程序工具、形式操作等尝试对传统建筑学进行变革与尝试。

在日益强调可持续的时代背景下，能量与热力学建筑的当代演化为建筑的自主性重建和广泛的社会响应提供了一个基于考古的未来视角。热力学法则指出每种自然形式的获得都是以最大化能量的供给与维持为原则，这种结构上对于能量的"捕获和引导"，成为创造新形式的重要契机。热力学建筑通过"物质—能量—形式—性能"之间的思考，强调了形式的开放性和介入建筑"大图景"的能力。

4. 陆家嘴

位于黄浦江东岸的浦东地区在 1990 年以前还是大面积的农田，零星有工业分布。进入 20 世纪 90 年代，上海政府决定扩大城市规模，在浦东陆家嘴地区进行了雄心勃勃的规划，打造具有国际水准的金融商务中心。如今，陆家嘴 CBD 的占地面积达到 6.8 平方公里，与曼哈顿下城金融区面积相当。陆家嘴贸易金融区在规划阶段就设想了由诸多标志性高层和超高层建筑来构成天际线，其中首个超高层建筑便是 1995 年完工的东方明珠电视塔。另外三座重要的标志性超高层建筑：金茂大厦、环球金融中心、上海中心也通过多轮国际竞赛相继建成，如今该区域内的高层、超高层建筑总量达到约 100 座，形成非常典型的高层、超高层建筑集群形态。

本次项目分为两个阶段：第一阶段针对空气与建筑、城市的关系做原型设计，第二阶段将原型植入具体的城市环境。基地位于上海陆家嘴金融贸易区，有着特殊的城市发展历史。我们的设计目标是针对该区域的空气污染、气流流动和建筑形态提出概念性的前沿设计。

5. 总结

本次工作营鼓励学生通过建造结构装置等方法来研究基于传统中国建筑的技术文化。同时，对中国当前的环境问题以及与此相关的建筑实践现象进行批判，总结出一套针对中国当前环境问题的新方法论。我们主张运用热力学与复杂科学的思维方法，重构中国当代城市建筑（作为研究对象）的边界问题、物质问题、能量流问题与组织形式问题，并对现代主义的建筑体系进行深刻的反思。

Appendix
The Summurise of Summer School

The 20th century was an era that people has created material civilization of unprecedented prosperity, but also an era that people produced significant damage to the earth's ecological environment and nature resources. The contradiction among city, building and environment has become increasingly serious, and it brought unexpected pressures to the development of the city at the same time. But with the growing concern on social issues, the autonomy of traditional architecture has faced a huge challenge, and architects have been fallen into a technical and theoretical ecology anxiety. In such a predicament, new transition of knowledge structure becoming more necessary.

2015 Summer School in CAUP is themed as "DESIGN AGAINST SMOG: Thermodynamic Methodology for Chinese Architecture", in which original ideas, methods and paradigms in response to the environmental crises in contemporary China will be explored and tested.

1. Smog

In 2015, the landmark documentary about China's catastrophic air pollution by former CCTV reporter Chai Jing has been circulated through social media and video sharing website. The documentary has sparked unprecedented concerns about the issue of smog. It forces architects, planners and engineers to come up with multi-disciplinary methods in response to the air pollution. Undoubtedly China is entering into an era which increasingly suffers environmental crises, reminiscent of the Los Angeles haze and London fog in the 1950s and 1960s. Paris, which still suffering from smog, trying to reduce the smog by the method of restricting the number of cars.

2. Air

What is air? A giant void nothingness? The transparent solvent of the world?

Where do we see air? Where do we encounter its force? The air ripples grass, scuds clouds, and crashes breakers against the shore.

The invisibility of air belies its volatile complexity, richness and density. A variety of particles can be manipulated into various textures, surfaces, and spaces.

Even when we feel that we are just breathing basic air, and are as "in the moment" as can be, we are actually inhaling a mosaic of airs that originated not just in various places, but at various historical times as well. In 2008, IN THE AIR is a visualization project which aims to make visible the microscopic and invisible agents of Madrid's air (gases, particles, pollen, diseases, etc.), to see how they perform, react and interact with the rest of the city.

Therefore, in this sustainable topic which appeals to improve climate and environment, the architects also have done a lot to try, they no longer only focus on the image and function of architecture, but pay attention to the climate, working on the space, air and flow, conduction, sweating, condensation and fluctuation in weather conditions, which will become the new contemporary architectural words. "What is atmosphere? Air or climate, or in between—effect, material, non-material, space and phenomena?"

Early in 1959, Buckminster Fuller's conception of a huge dome which located over Manhattan Island in New York, mapped out a sharp concern about environment in the high-speed period of modernization of western countries. In such an era, architecture will have to face the challenges of environmental crisis, and have to learn from Fuller's technical and cultural heritage.

3. Thermodynamics

Since Rudolf Clausius proposed the second law of thermodynamics in 1865, ENTROPY was introduced into the material as an important system state parameter. 100 years later, Ilya Prigogine created the complex scientific foundation of self-organization research on the dissipative system and the non-equilibrium system. According to Prigogine's point of view, building is an Open Non-Equilibrium System, and it is also a thermodynamic Dissipative Structure. This dissipative structure is characterized by the energy exchange and the entropy preservation, and must be considered as a whole thermodynamic system.

The invention of air conditioned building has often been designed as a closed micro-climate system based on fossil fuel consumption. Today, architects have to re-explore the flexible and open environment strategy of the traditional city—the microclimate of the architecture and the climate of the urban or regional as a whole, buildings adapt to and create a non-antagonistic internal micro-climate through active control the relationship of its external climate systems.

Thermodynamic architecture is trying to reconstruct a architectural critical and practice paradigm which bases on the energy flow and form generation. This is not just happening in the architecture, but also bases on the development of science and architecture in a century. It provides a new architecture type, form and pattern, integrates the body and tools, by the knowledge of interdisciplinary, discourse reconstruction and development tools. Peter Zumthor's museum in Bregenz, Austria, clearly showed the characteristics of the tectonics of building by the separation of the water cooling system and ventilation system in the wall; Philipp Lahm tried to present thermodynamics "encyclopedia" through a series of diagrams, and his artist's house vividly demonstrated the digestion of spatial boundary; Inaki Abalos, Kiel Moe, Sanford Kwinter et are launched the "Thermodynamic Architecture" experiment at Harvard University Graduate School of Design (GSD), through attempting to change the traditional architecture by knowledge system, programming tools and operation forms etc.

In this context of increasing emphasis on sustainable background, contemporary evolution of energy and thermodynamic architecture provides a future perspective on archeology for the autonomy reconstruction of architecture and widespread social response. Thermodynamic laws noted that each natural form is obtained in order to supply and maintain the principle of maximizing the energy. This structure of "capture and guide" the energy, is an important opportunity to create new forms. Thermodynamics architecture emphasizing the openness and intervention capability, by thinking of the "material, energy, forms and function."

4. Lujiazui

The project is divided into two steps: The first step is to do the prototype design against the relationship among air, architecture and city; The second step is to implant the prototype in the specific urban environment. The site is located in Lujiazui Finance & Trade Zone in Shanghai, it has a special history of urban development. Our design goal is to provide a conceptual design for the air pollution, the air flow and architectural form in this region.

The Pudong area, which located on the east bank of the Huangpu River, is a large area of farmland before 1990, where there is sporadic industrial distribution. In the 1990s, the Shanghai government decided to expand the size of the city, and carried out an ambitious plan to create a financial business center with international standards in the Lujiazui area. Today, that section, called Lujiazui, occupies an area of 6.8 kilometers, which is almost exactly the same shape and size as the financial district of Manhattan Downtown. Lujiazui Finance & Trade Zone had imagined by many landmark high-rise and super high-rise buildings to form the skyline in the planning stage. Three important landmark high-rise buildings: the Jinmao Tower, the World Financial Center and the Shanghai center also have been completed through several rounds of competition process. Now there are about 100 high-rise buildings in this area, forming a cluster of typical skyscrapers.

5. Conclusion

Therefore, the summer school encourage students to study in the Chinese traditional way to create microclimate (such as urban layout, architecture type, variable space, material practices, etc.), and study the technological culture based on the traditional Chinese architecture. We advocate using thermodynamics and complex scientific way, to reconstruct the China's contemporary urban's (as the object of study) border, material, energy flow and organization form issues, and deeply reflect on the architectural system of Modernism.

暑期学校指导及学生名单

List of Supervisors and Students

非"异形"热力学
指导：李麟学，马韬
学生：卡约·巴博萨，克里斯蒂安·拉维斯特，拉姆斯·支纪尧姆，路易斯·让，王立杨，杨之赟，钱韧

不可见的基础设施
指导：周渐佳，高军
学生：亚历山大·马蒂亚斯·雅各布森，艾尼斯·布洛东·博雷尔，詹姆斯·亚瑟·克莱夫·哈格雷福斯，张翔，张润泽，朱静宜

净化：生态介入
指导：谭峥，李卓
学生：艾德琳·康拉德，茱莉亚·埃索皮，杜杰，明磊，张博涵，吴潇

核：空气基础设施
指导：罗晶，王子耕
学生：庄田智己，刘芳铄，车进，武晓宇，梁芊荟，巴勃罗·马里亚诺·博纳·费尔南德斯 - 洛加

热动力呼吸系统
指导：吴迪，胡琛琛
学生：姜大宏，李鹜，吉姆·柏雷诺，姚慧婷，张帆，赵孔

藤
指导：邓闵衢，杨峰
学生：大卫·马萨里尼，维罗妮卡·佳佐拉，艾丽莎·玛丽斯黛拉，张振伟，温子申，孙童悦

退出雾霾
指导：苏运升，乔治·吉奥吉夫
学生：艾米丽娅·切卡·季米诺，伊万·A·韦弗，凌梦芷，周一凡，张程远，言语

循环的城市主义
指导：刘宇扬，高亦陶
学生：索菲亚·布兰科·桑托斯，宫本冬，姜信雄，皮欹，黄迪，毛宁俊

DeMONSTERative Thermodynamics
Supervisors：Li Linxue, Marta Pozo
Students：Caio Barboza, Christian Lavista, Ramus Guillaume, Louis Jean, Wang Liyang, Yang Zhiyun, Qian Ren

Invisible Infrasturcture
Supervisors：Zhou Jianjia, Gao Jun
Students：Alexander Matthias Jacobson, Inés Brotons Borrell, James Arthur Clive Hargrave, Zhang Xiang, Zhang Runze, Zhu Jingyi

Purify: Ecological Intervention
Supervisors：Tan Zheng, Li Zhuo
Students：Adeline Conrad, Giulia Esopi, Du Jie, Ming Lei, Zhang Bohan, Wu Xiao

Core: Air Infrastructure
Supervisors：Luo Jing, Wang Zigeng
Students：Tomoki Shoda, Liu Fangshuo, Che Jin, Wu Xiaoyu, Liang Qianhui, Pablo Mariano Bernar Fernandez-Roca

A Thermodynamically Driven Respiratory System
Supervisors：Lyla Wu Di, Hu Chenchen
Students：Kang Dahoon, Li Ao, Peraino Jim, Yao Huiting, Zhang Fan, Zhao Kong

The Vine
Supervisors：Minqu Michael Deng, Yang Feng
Students：Davide Masserini, Veronica Gazzola, Alyssa Maristela, Zhang Zhenwei, Wen Zishen, Sun Tongyue

STEP: Esc Smog
Supervisors：Su Yunsheng, Georgiev Georgi
Students：Amalia Checa Gimeno, Evan A Weaver, Ling Mengzhi, Zhou Yifan, Zhang Chengyuan, Yan Yu

Recy-Procal Urbanisms
Supervisors：Liu Yuyang, Oscar Ko
Students：Sofia Blanco Santos, Fuyu Miyamoto, Kang Shinwoong, Pi Qin, Huang Di, Mao Yujun

设计应对雾霾 DESIGN AGAINST SMOG

暑期学校活动场景
Summer School Activities

作者简介

伊纳吉·阿巴罗斯

1956年出生于西班牙圣塞巴斯蒂安，哈佛大学设计研究生院建筑系主任，建筑系教授，阿巴罗斯＋森吉维克建筑事务所主持建筑师。1978年获马德里理工大学建筑学硕士学位，1991年获博士学位。他曾任纽约哥伦比亚大学、伦敦建筑联盟、洛桑联邦理工大学及康奈尔大学访问教授，是普林斯大学让·拉巴图特教席教授，哈佛大学丹下健三教席教授。1984年至2007年，伊纳吉·阿巴罗斯与胡安·埃莱罗斯共同成立建筑事务所，2007年独立执业，之后与雷娜塔·森吉维克成立"阿巴罗斯+森吉维克"建筑事务所。他的建筑作品曾经被众多知名建筑杂志广泛报道。曾出版专著《柯布西耶摩天大楼》(1988)，与埃莱罗斯合著包括《塔楼与办公室》(2003)及《自然人工》(1999)等多本重要作品。自2005年任西班牙建筑 — 城市 — 能源国际大会科学委员会成员。2014年担任第14届威尼斯建筑双年展西班牙馆策展人。

———

李麟学

同济大学建筑与城市规划学院教授，博士生导师，麟和建筑工作室ATELIER L+主持建筑师，能量与热力学建筑中心CETA与社会生态实验室SOCIOECO LAB主持人，哈佛大学GSD设计研究生院高级访问学者(2014)，《时代建筑》专栏主持人。2000年曾入选法国总统交流项目"50位建筑师在法国"，在巴黎建筑学院PARIS-BELLEVILLE学习交流。李麟学试图以明确的理论话语，确立建筑教学、研究、实践与国际交流的基础，将建筑学领域的"知识生产"与"建筑生产"贯通一体。主要的研究领域包括：热力学生态建筑、公共建筑集群、以及当代建筑实践前沿等。基于热力学生态系统研究，发掘能量与建筑本体互动的新范式，以此确立教学与研究的主线。基于其"自然系统建构"的建筑哲学与创造性实践，成为中国当代建筑的出色诠释者之一，也是国际学术领域热力学建筑与生态公共建筑集群的积极推动者。

李麟学主持建成杭州市民中心(2013世界高层建筑学会"世界最佳高层建筑"亚太区提名奖)、2010中国上海世博会城市最佳实践区B3馆、四川国际网球中心、中国商贸博物馆等多项有影响力的建筑作品。曾获得中国建筑学会"青年建筑师奖"(2006)，上海青年建筑师"新秀奖"(2005)等荣誉，获得国内外各类专业设计奖项二十余项。曾参加40位小于四十岁的华人建筑师设计作品展、2010中法建筑与城市发展论坛、中国比利时青年建筑师交流上海论坛、"从研究到实践"米兰建筑三年展(2012)、威尼斯建筑学院国际工作室特邀教授(2013)、上海西岸当代建筑与艺术双年展(2013)、深港城市建筑双城双年展(2013)、上海城市空间艺术季城市更新展(2015)、"走向批判的实用主义：当代中国建筑"(哈佛GSD,2016)等展览、论坛与学术活动。主持国家自然科学基金资助项目"基于生态化模拟的城市高层建筑综合体被动式设计体系研究""能量与热力学建筑前沿理论建构"等重要课题，在国内外核心专业刊物发表论文四十余篇，客座主编时代建筑《形式追随能量：热力学作为建筑设计的引擎》。个人及作品曾入选《建造革命：1980年以来的中国建筑》《中国年轻一代的建筑实践》等专著。

———

周渐佳

上海冶是建筑工作室创始合伙人、主持建筑师，《时代建筑》杂志栏目主持人，同济大学建筑与城市规划学院客座讲师，香港大学建筑系上海中心讲师。

———

王子耕

普林斯顿大学建筑学硕士，北京电影学院美术学院教师，PILLS主编。曾获得普林斯顿大学全额入学奖学金资助，FAIRYTALES 2015国际竞赛一等奖，中国建筑设计研究院城市规划专业方案设计一等奖，建筑专业方案设计一等奖等。作品参展威尼斯国际建筑展，大声展及港深双年展等重要国内外展览，并在多所大学担任客座教师和评委工作。

艾奥利奥索托·汉德拉纳塔

被誉为索托，毕业于英国卡迪夫大学威尔士学院和美国麻省理工学院。就职于迪勒斯科菲迪奥驻纽约办公室，负责在欧洲和美国的公民和文化项目。最近，他开始与蒂森·希尔曼合作创立研究组织ma-en（www.ma-en.com），关注并研究市民和公共空间的形式表达。2013年，他荣获KPF建筑事务所旅行奖学金，展开对"建筑，工具和环境"的研究。同时他也是SOM建筑设计事务所2014年旅行奖学金的获得者，他的研究基于灾后重建的视角，探讨"建筑与灾难之间的关系"。这两项研究，与他的硕士论文相结合，形成了基于文化的独特的建筑和城市设计实践主题。

谭峥

同济大学助理教授，张永和教授"千人团队"教学科研执行人。获加州大学洛杉矶分校建筑学博士学位，为该系首位华人博士学位获得者。长期研究与基础设施相关的城市形态与建筑策划，在国内外期刊发表大量关于城市建筑学的文章，同时任《时代建筑》客座编辑与《新建筑》杂志审稿人。

刘宇扬

出生于台湾并成长于美国，刘宇扬先生拥有近二十年的海内外专业及学术背景。刘先生毕业于美国哈佛大学设计学院，师从荷兰建筑家雷姆·库哈斯，完成中国珠江三角洲城市化的研究，受邀参与德国卡塞尔第十届文件展，并共同出版 Great Leap Forward 一书。刘先生自1997年起在北美、香港和上海长期从事建筑实践、研究和教育工作。刘先生除了领导事务所的日常运营及管理工作外，目前还担任上海建筑学会学术创作委员，上海市青浦区规土局顾问建筑师，及《Domus国际中文版》嫩鸟计划终身导师等公益性职务，并长期受邀参与国内外院校和专业领域的教学、讲座、及策展活动。

苏运升

城市规划博士，跨学科跨尺度智能城市策划规划专家城市智能模块产品和服务创新设计师，全程参与上海2010年世博会的规划和设计，央视英语频道世博会开闭幕式主持嘉宾，主持过广州金融城，福州三江口等多个国内中心区城市设计以及安哥拉、赞比亚、俄罗斯、越南等多个国际城市的规划项目。2014在全球TED、腾讯WE大会，央视一人一世界节目等场合主讲未来智能城市创新。多次受邀在上海浦东干部培训班及国家行政学院市长局长培训班授课，并担任同济大学创新孵化器以及多家机构及公司的创新顾问以及导师。

乔治·吉奥吉夫

毕业于亚琛工业大学并获得城市规划科学硕士学位，是斯图加特大学轻型建筑及概念设计研究所的助理教授。自2014年以来，从事关于可持续建筑和规划的研究与教学。同时，他也是弗劳恩霍夫建筑物理研究所研究项目（2013）的主要负责人，聚焦于基于智能能源效率优化的多元产权建筑和复合空间的技术和协调过程研究。

李卓

博士，副教授，于2007年获得西安交通大学工程热物理博士学位，2010年获得英国曼彻斯特大学化学工程博士学位。在英国留学期间，先后获得两次科学研究的奖项。随后，前往巴黎高等物理化学学校继续研究工作。于2013年秋季回国，加入同济大学环境科学与工程学院。目前主要研究方向包括城市环境中大气污染物的扩散机理，室内环境污染物的传质机理以及开发新型技术用于检测饮用水中有机污染物。

设计应对雾霾 DESIGN AGAINST SMOG

陶文铨

西安交通大学、同济大学教授、博士生导师，中国科学院资深院士。陶文铨教授在传热与流动的先进数值计算方法及其应用、强化传热的基本理论与工程应用方面做出了重要的贡献。目前，陶教授的主要研究方向集中在城市地区污染物流动扩散的多尺度、跨尺度数值模拟与风洞实验研究，建筑室内挥发有机物 (VOC)/半挥发有机物 (SVOC) 及其复合污染传质机理研究。陶文铨教授现任多本著名国际传热传质领域期刊编委和 ICHMT 理事会委员。目前，陶教授已发表国际期刊文章 500 余篇，做特邀或主题报告 40 余次，获 2004 国家自然科学奖二等奖和 2009 国家技术发明奖二等奖各一项，发起了"亚洲计算传热学学术会议 (ASCHT)"，编著著作 10 余部，所著专著《数值传热学》已经被国内外期刊论文引用六千余次。

About Authors

University in 2014, and chairman of special column in *T+A*. In 2000, he was selected by the Presidential Program "50 ARCHITECTES EN FRANCE" and studied in Ecole d'Architecture de Paris-Belleville. Li Linxue tries to establish his architectural teaching, research, practice and international exchange based on the definite theoretical discourse and to integrate the production of knowledge and production of buildings in the field of architecture. His main fields of researches include thermodynamic ecological architecture, public architectural conglomerate, and frontier for contemporary architectural practice, ect. Basing on the studying for thermodynamic ecological system, exploring for the new paradigms of interaction between energy and architectural noumenon, the main stream of teaching and research is created. He has the honor to become one of the excellent interpreters for contemporary Chinese architecture through his "nature based system" architectural philosophy and creative practice. He is also an active promoter for thermodynamic architecture and ecological public architectural conglomerates in the international academic field. Li Linxue is the responsible architect for many influential projects such as Civic Center of Hangzhou, which is 2013 Best Tall building Nominee of Asia & Australasia Region from CTBUH, Hall B-3 in Urban Best Practice Area of EXPO 2010 Shanghai China, Sichuan International Tennis Center, and China Commerce and Trade Museum, etc. He is honored with Young Architect Award (2006) by Architectural Society of China, Shanghai Young Architect Rookie Award (2005), and more than twenty international and domestic professional design prizes. He attended 40 Under 40 Exhibition, 2010 Sino-France Architecture and Urban Development Forum, Shanghai-Brussels ARCHITopia Forum, from Research to Practice (Milan Triennal, 2012), IUAV Workshop as a guest professor (2013), West Bund Biennale of Shanghai, Bi-city Biennial of Urbanism/ Architecture-Shenzhen (2013), Shanghai Urban Space Art Season-Urban Regeneration (2015), and Towards A Critical Pragmatism: Contemporary Architecture in China (GSD Harvard, 2016), etc. Meanwhile he has charged important programs like the research of Passive Design System for Urban High-rise Building Complex Based on the Ecological Simulation financed by National Natural Science Foundation of China, and Construction of the Advanced Theory for Energy & Thermodynamic Architecture, ect. He's authored more than forty papers in both international and national core journals, and coedited *Form Follows Energy: Thermodynamics as The Engine of Architectural Design* for *T+A*. His works are selected in the books like *Building A Revolution: Chinese Architecture Since 1980*, *Architectural Practice of Chinese Young Generation*, ect.

Iñaki Abalos
Iñaki Abalos is a Ph.D. in Architecture (1991) and Chaired Professor of Architectural Design at the ETSAM (since 2002). He was Kenzo Tange Professor (2009), Design Critic in Architecture (2010-2012), and since 2013 Professor in Residence and Chair of the Department of Architecture at Harvard University Graduate School of Design. He has been Visiting Professor in Columbia University (New York), Architectural Association (London), EPFL (Lausanne), Princeton University (New Jersey) and Cornell University (Ithaca). In association with Renata Sentkiewicz, he is a founding member of Abalos+Sentkiewicz since 2006. The work they develop stands out for proposing an original synthesis of technical rigor, formal imagination and discipline integration between architecture, environment and landscape, an approach they have named "A Thermodynamic Beauty".

Li Linxue
Prof. Dr. Li Linxue is a Ph.D. Supervisor at College of Architecture and Urban Planning of Tongji University, principal architect of ATELIER L+, director of CETA (Center for Energy & Thermodynamic Architecture) & SOCIOECO LAB, visiting scholar at Graduate School of Design Harvard

Zhou Jianjia
Zhou Jianjia is the co-founder and principal architect of YeArch Studio, co-editor of *T+A* Magazine, guest lecturer of College of Architecture and Urban Planning, Tongji University and Hong Kong University Shanghai Study Center.

Wang Zigeng
Princeton University, is the founder of the Studio of the Speculative Society and the joint initiator of the DA League. He once worked for FCJZ, Atelier Li Xinggang of China Architecture Design & Research Group and Z-studio of Zhubo Design Group Co., Ltd. He also worked as AI and visiting critic at Princeton University, Tongji University and the China Central Academy of Fine Arts. He won the Excellence Award Annual National Architectural Design Competition for University Students four times, and the first prize for the FAIRYTALES 2015.

Erioseto Hendranata
Erioseto Hendranata, known as Seto, trained at both the Welsh School of Architecture, Cardiff University and at Massachusetts Institute of Technology (MIT). He works as a designer at Diller Scofidio & Renfro (DS+R) office in New York on their civic and cultural projects in Europe and

America. Recently, he started a collaboration called ma-en (www.ma-en.com) with Dessen Hillman with main interest in the expressions of civic and communal spaces. In 2013, he was awarded the Kohn Pedersen Fox (KPF) Traveling Fellowship to conduct research on "Buildings as Instruments and Environments". Seto was the winner of Skidmore Owings Merrill (SOM) Prize 2014. His research investigates the relationship between architecture and crises - focusing on the post-disaster reconstruction effort. Both of these research, combined with his Master thesis, have formed certain themes of enquiries on the culturally unique architecture and urban design practices.

Tan Zheng
Tan Zheng is Assistant Professor in architecture and urbanism at Tongji University. As an urban designer and architectural historian, Tan has taught at Los Angeles, Hong Kong and Shanghai. His research revolves around contemporary urban form with a focus on infrastructures and interiorized public spaces. He is currently writing a book on the emerging urban code in Shanghai's new neighborhood design. Tan earned his Ph.D. in Architecture at the University of California, Los Angeles. His dissertation investigates Hong Kong's civic spaces in relation with high density, consumerism and public transportation. His previous researches and creative works include design methodology, digital visualization and architectural programming.

Liu Yuyang
Born in Taiwan, Mr. Liu Yuyang received his Master in Architecture from Harvard Graduate School of Design and his B.A. from the University of California, San Diego. While at Harvard, Mr. Liu researched with Rem Koolhaas and co-authored *Great Leap Forward*, a seminal work on the urbanization of China's Pearl River Delta. Having held a number of professional and academic positions in the US, Hong Kong and Shanghai since 1997. Mr. Liu now leads his Atelier Liu Yuyang Architects, a Shanghai based critical practice, and also serves as a member of the Shanghai Architectural Society Academic Committee, an advisory architect for the Shanghai Qingpu District Planning Bureau, and has been teaching, lecturing and exhibiting frequently home and aboard.

Su Yunsheng
Dr. Su Yunsheng is the Innovation Director of Master Planner for the Shanghai 2010 World Expo., a guest lecturer at Shanghai's Urban Planning Bureau, and co-founder of *Urban China* Magazine. He has given lectures at Tongji University in Shanghai, AA school in UK, Bauhaus School as well as Technische Universität Darmstadtin in German, etc. Dr. Su earned his bachelor's, master's & doctor's degree all in Urban Planning, Design and Design Theories from Tongji University. His career at Shanghai Tongji Urban Planning & Design Institute started in 2001. He has supervised many important Urban Design projects, such as Beijing Tongzhou Canal city, 2010 Shanghai Expo-village, 2013 Guangzhou Financial City etc. He has been active in bringing the experience of China's rapid modernization and its designs to other developing countries. In 2005, he undertook the task of promoting Tongji University globally and introduced the concept of "China Cities". The same year, Dr. Su played a role in launching Urban China Magazine, a modern planning and design publication. In 2008, together with some other Tongji professors, he found Etopia, a Building design & construction platform, aiming at innovation for life.

Georgi Georgiev
Georgi Georgiev, who earned a Master of Science of Urban Planning at RWTH Aachen, is an assistant professor of Institute for Lightweight Constructions and Concept Design ILEK of University Stuttgart. He has been researching and teaching in the area of sustainable building and planning since 2014. Also he has been a research project manager in Fraunhofer-Institute for Building Physics IBP since 2013, focusing on the area "Technics and negotiation process in multy-ownership building and spatial complexes for smart energy efficiency optimization".

Li Zhuo
Dr.Li Zhuo has been appointed as an associate professor in the State Key Laboratory of Pollution Control and Resource Reuse of Tongji University since 2013. She got her first Ph.D. in engineering thermophysics at Xi'an Jiaotong University in 2007. Afterwards, she moved to Britain and got her Ph.D. in chemical engineering at the University of Manchester in 2010. During her time in U.K., she won the 5th British Biology Physics Conference Prize at Imperial College London in 2007 and the BSR-TA prize at the University of Edinburg in 2009. From 2010 to 2013, she worked as a research fellow in École Supérieure de Physique et de Chimie Industrielles (ESPCI), Paris. Dr. Li has published more than 20 articles in the journals such as *Advanced Functional Materials, Solar Energy Materials and Solar Cells, Solar Energy, International Heat and Mass Transfer, Microfluidic Nanofluidic* etc. At present, she mainly concentrates on the studies of the flow and dispersion of pollutions in the urban area based on the computational fluid dynamics (CFD)and wind-tunnel experiment, the mechanisms of the emissions and transport of volatile or semivolatile organic compounds (VOCs/SVOCs) in indoor environments, and develops new techniques in monitoring the disinfection by-products (DBPs) in drinking water.

Tao Wenquan
Prof. Tao Wenquan is a full professor at Xi'an Jiaotong University, a guest professor at College of Environmental Science & Engineering of Tongji University, and a member of Chinese Academy Science. Prof. Tao made remarkable contributions in the advanced numerical method and application in heat and mass transfer, the heat enhanced theory, mechanisms and engineering applications. Recently, Prof. Tao has focused his studies on the advanced multi-scale numerical method of pollutants transport in urban environments, and the mass transfer mechanisms of VOC/SVOC in indoor environments. Professor Tao serves as an editor of Journal of Xi'an Jiaotong University (Natural Science Ed.) and associate editor of International Journal of Heat & Mass Transfer, International Communications in Heat & Mass transfer and ASME Journal of Heat Transfer. He is also a member of advisory board of Numerical Heat Transfer and editorial board of Progress in Computational Fluid Dynamics, and a member of scientific council of International Center for Heat and Mass Transfer. Prof. Tao has published more than 500 papers in international journals, and has been invited to give theme reports for more than 40 times. He won the second prize of National Natural Science 2004, and National Technology Invention Award 2009 every one time. He launched the "Asian Symposium on Computational Heat Transfer and Fluid Flow (ASCHT)", and compiled over 10 books. His monograph—*Numerical Heat Transfer* has been cited more than six thousand times at home and abroad.

设计应对雾霾 DESIGN AGAINST SMOG

主编简介

Introduction of Chief Editor

李麟学

Li Linxue

同济大学建筑与城市规划学院教授、博士生导师，麟和建筑工作室 ATELIER L+ 主持建筑师，能量与热力学建筑中心 CETA 与社会生态实验室 SOCIOECO LAB 主持人，哈佛大学 GSD 设计研究生院高级访问学者（2014），《时代建筑》专栏主持人。2000 年曾入选法国总统交流项目"50 位建筑师在法国"，在巴黎建筑学院 PARIS-BELLEVILLE 学习交流。李麟学试图以明确的理论话语，确立建筑教学、研究、实践与国际交流的基础，将建筑学领域的"知识生产"与"建筑生产"贯通一体。主要的研究领域包括：热力学生态建筑、公共建筑集群，以及当代建筑实践前沿等。李麟学主持建成杭州市民中心（2013 世界高层建筑学会"世界最佳高层建筑"亚太区提名奖）、2010 中国上海世博会城市最佳实践区 B3 馆、四川国际网球中心、中国商贸博物馆等多项有影响力的建筑作品。曾获得中国建筑学会"青年建筑师奖"（2006）、上海青年建筑师"新秀奖"（2005）等荣誉，获得国内外各类专业设计奖项二十余项。曾参加 40 位小于四十岁的华人建筑师设计作品展、"从研究到实践"米兰建筑三年展（2012）、威尼斯建筑学院国际工作室特邀教授（2013）、深港城市建筑双城双年展（2013）、上海城市空间艺术季城市更新展（2015）、"走向批判的实用主义：当代中国建筑"（哈佛 GSD，2016）等展览、论坛与学术活动。主持国家自然科学基金资助项目"基于生态化模拟的城市高层建筑综合体被动式设计体系研究""能量与热力学建筑前沿理论建构"等重要课题，在国内外核心专业刊物发表论文四十余篇，客座主编时代建筑《形式追随能量：热力学作为建筑设计的引擎》。基于其"自然系统建构"的建筑哲学与创造性实践，成为中国当代建筑的出色诠释者之一，也是国际学术领域热力学建筑与生态公共建筑集群的积极推动者。

Prof. Dr. Li Linxue is a Ph.D. Supervisor at College of Architecture and Urban Planning of Tongji University, principal architect of ATELIER L+, director of CETA (Center for Energy & Thermodynamic Architecture) & SOCIOECO LAB, visiting scholar at Graduate School of Design Harvard University in 2014, and chairman of special column in T+A. In 2000, he was selected by the Presidential Program "50 ARCHITECTES EN FRANCE" and studied in Ecole d'Architecture de Paris-Belleville. Li Linxue tries to establish his architectural teaching, research, practice and international exchange based on the definite theoretical discourse and to integrate the production of knowledge and production of buildings in the field of architecture. His main fields of researches include thermodynamic ecological architecture, public architectural conglomerate, and frontier for contemporary architectural practice, ect. Li Linxue is the responsible architect for many influential projects such as Civic Center of Hangzhou which is 2013 Best Tall building Nominee of Asia & Australasia Region from CTBUH, Hall B-3 in Urban Best Practice Area of EXPO 2010 Shanghai China, Sichuan International Tennis Center, and China Commerce and Trade Museum, etc. He is honored with Young Architect Award (2006) by Architectural Society of China, Shanghai Young Architect Rookie Award (2005), and more than twenty international and domestic professional design prizes. He attended 40 Under 40 Exhibition, from Research to Practice (Milan Triennal, 2012), IUAV Workshop as a guest professor (2013), Bi-city Biennial of Urbanism/ Architecture-Shenzhen (2013), Shanghai Urban Space Art Season-Urban Regeneration (2015), and Towards A Critical Pragmatism: Contemporary Architecture in China (GSD Harvard, 2016), etc. Meanwhile he has charged important programs like the research of Passive Design System for Urban High-rise Building Complex Based on the Ecological Simulation financed by National Natural Science Foundation of China, and Construction of the Advanced Theory for Energy & Thermodynamic Architecture, ect. He's authored more than forty papers in both international and national core journals, and coedited *Form Follows Energy: Thermodynamics as The Engine of Architectural Design* for T+A. Li Linxue has the honor to become one of the excellent interpreters for contemporary Chinese architecture through his "nature based system" architectural philosophy and creative practice. He is also an active promoter for thermodynamic architecture and ecological public architectural conglomerates in the international academic field.